Anthrophysical Form

Anthrophysical Form

Two Families and
Their Neighborhood Environments

Robert L. Vickery, Jr.

University Press of Virginia

Charlottesville

The University Press of Virginia
Copyright © 1972 by the Rector and Visitors
of the University of Virginia

First published 1972

ISBN: 0–8139–0393–9
Library of Congress Catalog Card Number: 73–183896
Printed in the United States of America

For my mother

Contents

Foreword

FOR some time now architects have been looking earnestly at groups of buildings and whole settlements that have no pretensions to high architectural art and offer no significant technological innovations in the practical science of building. In contrast to the past, when they concentrated on important monumental buildings, temples, churches, and palaces, architects now seem increasingly caught up in the everyday environments in which most people spend their lives.

Many recent books, articles, and published collections of pictures attest to the new interest. The present study joins these, documenting a concern shared by that generation of younger architects just beginning to take over responsible leadership in practice and in the schools of design. But this book goes beyond many architectural studies of everyday living environments by examining more than the physical layouts and the census statistics. It looks intimately at family life as well, focusing closely on the people inhabiting the places studied and on their personal lives there. Other efforts in this genre, for instance *Urban Dwelling Environments* by Horacio Caminos, John Turner, and John Steffian, offer land-use and demographic data along with the handsome graphics conventionally provided by architectural authors. In contrast, Vickery's *Anthrophysical Form* gives a much more concrete sense of the life actually lived in the settings it describes.

One thing may trouble readers about Vickery's turn toward an anthropological method of study and reporting—the isolation of his work from the growing list of studies that apply social and behavioral science to the explication of everyday physical environments. Academic considerations would suggest that the influence of Hall, Michelson, Perin, and Craik might have enriched the study. But this may not have been possible. The study of everyday physical environment from a perspective that joins architecture with anthropology, psychology, and sociology has only become possible in the same years Vickery was studying St. Louis and Ahmedabad. Theoretical interpretation does not easily fit together with field work such as that described in this book. Clearly the author did not set out to do both, to build social theory *and* to describe physical facts.

His concern with description leads to one of the signal strengths Vickery shares with others among the new architect-environmentalists: their concern with form itself. *Anthrophysical Form* deals with the spatial patterns and en-

vironmental space, the morphology, of everyday life in a St. Louis private
street subdivision and in a village near Ahmedabad. It presents these forms in
carefully measured scale drawings, the special language architects know best.
This reverses the usual situation in which no graphics, or at best a few dim
snapshot pictures, merely illustrate a text presented in verbal and numerical
symbols. Vickery provides a graphic document supported with verbal and
some numerical illustrations. In doing so he demonstrates that the grand tradition of fine architectural draftsmanship is very much alive and relevant, even
though the draftsman's subject matter has changed radically—from the classical
orders to mud huts.

Turning to the specific places discussed in *Anthrophysical Form,* the careful
documentation of the St. Louis private street phenomenon helps fill a long-
standing gap. Few American contributions to urban design have gone so un-
remarked, and yet have counted for so much in the evolution of settlement
design in the twentieth century. Vickery offers tantalizing hints on the un-
studied connections between St. Louis's remarkable Major Pitzman, through
Henry Wright, to the New York-based group including Clarence Stein and
Clarence Perry who gave the world the neighborhood unit concept. The graphic
documentation of Parkview Place in St. Louis makes the morphological case
for the connections irrefutable.

A final aspect of the study is that it offers a clue to the ferment in the St.
Louis architectural world during the 1950s and 1960s, when Robert Vickery
was first a student and then a member of the distinguished faculty of the
School of Architecture at Washington University. In those years, at that school,
a remarkable transformation occurred. The place became a favored outpost
of the new architectural movements in Europe, Asia, and the rest of the United
States, a place where the phenomenology of everyday human settlements replaced the monuments of elite culture as the chief subjects of architectural
interests.

In this sense Robert Vickery's *Anthrophysical Form* symbolizes a good deal
more than what appears on the surface. It documents the shift of architecture
away from what C. Wright Mills derided as "beautifying the isolated *milieux*
of wealthy persons" toward a concern with "the planning of the human landscape for all people."

ROGER MONTGOMERY

Berkeley, California
June 1971

Preface

As is typical of most research, the investigations which led to this book began with a number of biases—among them the belief that although present urban housing environment was deteriorating, the cities were still worth saving and living in; that young families with children and economic mobility had to be induced to stay if the cities were to remain viable, and that, for the children, grass, trees, air, and places to play were critical; that American urban housing since 1945 had not, in any effective manner, approached this problem, and had furthermore been formally inappropriate and aesthetically vulgar; and finally, that most of what we thought of as suburbia would be an urban slum by the year 2000.

Almost ten years later I am no longer convinced that the city can in fact be saved.[1] Whether one can even begin to discuss the physical rebuilding of urban housing environment without first resolving issues of jobs, schools, integration, and fear, is debatable. But these issues are of immense national and local political significance, and while architects and planners must share in the concern, the solutions have to involve sociologists, lawyers, economists —indeed every citizen of the city. (Architects, however, should not dabble in sociology, and society should stop expecting them to). Nevertheless, I am more certain than ever that if saving the city is possible, then neighborhood form will be a critical issue, and that the architect-planner's unique role will be in determining the shape of this form. To do this, he will need tools—the best legal, tax, and census data, sociological surveys, economic feasibility studies, the latest construction technology. Even more important, however, he will need methodologies for research into what constitutes good, or aesthetically pleasing, environment.

This study is the final result of an attempt to discover and outline one such methodology. I have called it an anthrophysical approach to housing, because it utilizes certain anthropological interview techniques to investigate physical form. This approach is fully defined and explained in the first chapter. The second and third chapters consist of two case reports in which the method is applied to specific environments, and the final chapter presents limited conclusions and comparisons between the two case reports and puts forward a few

1. I am referring to the city in its traditional sense (Florence, for example) as an integrated entity of culture, commerce, housing, and professional business. The development of Los Angeles leads one to suspect the city may be moving toward a grouping of subcenters linked by rapid transit in which the old core becomes an open park, or a cultural repository, or a subcenter itself composed of militant blacks (and probably surrounded by real walls).

proposals for future housing study. I have never intended the research methodology, or the book, to be conclusive. They are intentionally open-ended, hopefully suggesting to architects and planners a beginning rooted in anthrophysical concerns.

In preparing this work I have become indebted to many persons in the academic world for discussion and advice. Among them are Joseph Passonneau, former dean of the Architecture School at Washington University in St. Louis, and professors Roger Montgomery, Fumihiko Maki, and Roland Bockhorst. In fact, most of the architectural faculty at Washington University was at some time helpful, as has been the faculty of the School of Architecture at the University of Virginia. At the latter institution I am particularly grateful to Dean Joseph Bosserman, and professors Frederick Nichols, Thomas Czarnowski, Simon Pepper, and Andrew Sammataro.

Individuals to whom I am indebted are Mrs. Mildred Scott (for intellectual insights far more than for typing); Mr. Lawrence Stern, for introductions in Parkview; Mr. B. V. Doshi, for introductions in India; and Mr. Craig Eppes, for continual encouragement.

A number of my students have also been instrumental, particularly those who have worked on sketches and drawings—these include Mr. Tom Eckelman, Mr. Masao Yamada, Miss Gita Shah, and Miss Gay Goldman. In particular I thank Mr. George Miers, who prepared the vast majority of the final drawings used, as well as assisting in rethinking certain portions of the text.

For financial assistance I must mention the Steedman Foundation in St. Louis (for travel funds), the American Institute of Architects (for a grant to study housing), Washington University (for support in preparing the drawings), and the National Endowment for the Arts in Washington, D.C., a federal agency created by Act of Congress in 1965, which supplied a substantial grant to complete final writing and editing.

Finally I must thank my close friends and immediate family, and in particular my father, and my wife, who never gave up hope that I *would*, someday, finish.

ROBERT L. VICKERY, JR.

Charlottesville, Virginia
September 1971

Anthrophysical Form

I   An Anthrophysical Approach
to Neighborhood Environment

For a long time I had been fascinated by the unique private street systems of St. Louis, and in 1967 I decided to begin a modest investigation into why one of them, Parkview Place, was a successful environment and a pleasant area in which to live. This neighborhood attracted young, professionally oriented families with children, and I felt that keeping such potential civic leaders within the urban core was necessary if the city were to remain viable. What began as a modest study soon grew into a chaotic collection of facts, opinions, maps, drawings, photographs, social surveys, and historical data. As the accumulation continued, two obvious problems in research methodology emerged.

The first was one of definition. I found that in questioning what constituted good neighborhood form, one first had to determine what constituted good architecture, and in so doing, to redefine the nature of architectural aesthetics. This is now somewhat old ground. Most of us today recognize that architecture can no longer be thought of solely in the poetic terms of "masses in light," or "frozen music" or the "mother of the arts," any more than it can be thought of simply as a technical approach to the profession of building.

The confused state of our aesthetics, brought on both by the sweeping changes in our social philosophy and the rapid advances in our construction technology, demands that we return to basic principles, recognizing architecture as that discipline of art concerned with the shaping of physical environment.

While this definition may rule out pollution, population control, and integration, thus limiting the role of the architect-planner, it in no way demeans his task. The physical designer must see the city as a four-dimensional, linked sequence of events in which human activities are concentrated at the nodal points. (Aldo Van Eyck, the Dutch architect, calls this concept the marriage of "place with occasion.") This perspective requires not only a concern for one's society and its aspirations, but also the ability to translate these aspirations creatively into shaped form at many levels of visual perception.

To investigate whether the Parkview Place neighborhood "worked aesthetically"—that is, to discover whether the physical environment enhanced the lives of the residents—it became necessary to turn to basic questions. For whom was the neighborhood designed? Who lives there now? What are their real needs? Does the environment help satisfy these needs? What elements of form are used to shape the environment? Could these elements be reshaped using new technology to improve the environment? From the resolution of these questions the criteria could arise for developing a new aesthetic in housing design. This concept is nicely stated by Walter Schwagenscheidt in *A Wanderer in the City:*

Man does not live from bread alone, but requires other nourishment from childhood on. To give house and city the correct form, any preconceived ideas must be abandoned at the start. One house is not like another, nor is one city like another. There is no "must" in the design of a house. There is no "must" in design at all. All sorts of factors have to be considered. The secret is to let the factors speak for themselves, and not to keep babbling as though we knew better than the factors themselves.[1]

The second problem in research methodology was in organization. The sources of information proved to be endless, and the quantity of data for making any decisions as to what was relevant in understanding physical form seemed overwhelming. Professor Roger Montgomery, then teaching at Washington University, suggested a solution to this dilemma: if rather than attempting to organize statistically a great mass of data (for which sociologists should be better trained than architects), one instead selected a controlled group of people and studied in depth how they responded to the environment, he could begin to determine which elements of form responded to their needs. The important aspects of this approach were that its selectivity emphasized a study of physical form. This would make the research conclusions particularly interesting to architects and planners. It also meant that an architect could do the research himself rather than having to rely on a team of specialists.

Oscar Lewis's *Five Families in Mexico* (New York: Basic Books, 1969) provided a final organizational clue. Lewis's book concerns the ways in which families of different backgrounds respond to the *social* environment of their neighborhoods. It is thus an anthropological book, with architectural implications. I reasoned that if I were to study in depth how one family in Parkview Place responded to the *physical* environment of its neighborhood, I could then do architectural research, with anthropological implications.

The anthrophysical approach to understanding neighborhood form is in fact a simple method, utilizing three stages of investigation. The first is a physical inventory or description of the environment, with emphasis on the determinants

1. Walter Schwagenscheidt, *A Wanderer in the City* (Berlin: Hans Bruntsch and Co., 1957), p. 6.

of form. The second is a study in depth of how one family interacts with its environment—how it uses its neighborhood. The third stage is a consideration of how future housing might be improved through an understanding of the issues discovered during the study.

The approach is termed *anthrophysical* to emphasize the contention that physical design cannot exist independently of human need. While planners must be concerned with form, they must also remember that only as form is shaped to satisfy and delight each and every man does it become architecture.[2] The research must therefore be conducted by someone trained in physical design.

First Stage: The Description of the Environment

The first stage involves an investigation into the historical evolution of a particular neighborhood form, utilizing newspaper accounts, old plat plans, and photos in an attempt to determine the cultural orientation of the original builders. The investigation leads naturally into a physical description of the present environment, with emphasis on the determinants of form.[3] These factors will vary from one neighborhood to another, but among the more important are manifestations of privacy (such as setbacks, blank walls, shrubbery, fences, siting); elements of continuity (in materials, scale, sidewalks, tree planting, color); and means of punctuation or variety (such as gates, towers, monuments, unexpected vistas, parks).

Certain drawings are found to be critical at this stage. First, a base map should always be drawn, reflecting houses, circulation systems, and special planting. Two overlays to this drawing are essential: one showing the circulation movement (intensity of use); and another showing public and private spaces. (A correlation has always been found to exist between these two overlays.) In addition, other overlays may be used when it is felt they would add to an understanding of the form (for example green space vs. paved space, or enclosed space vs. open space). To these map overlays may be added other drawings, such as sections, plot plans, and sequential movement diagrams, depending on the direction of the research.

2. This relates to humanism in architecture, and by implication questions much of the current educational practice. The role of the architect-planner must be as a philosopher of future form for humans (as opposed to future form for automobiles, sewers, lighting systems, and industrial parks). The inability of many planners to do anything more than color zoning maps (when they are not busy running new arterial streets through parks), is deeply disturbing.

3. A prior familiarity with these architectural determinants is essential. For those not skilled in recognizing these elements, I recommend *Image of the City* by Kevin Lynch (Cambridge, Mass.: Harvard University Press, 1960); *Community and Privacy*, by Serge Chermayeff and Christopher Alexander (Garden City, N.Y.: Doubleday and Co., 1963); *Townscape* by Gordon Cullen (New York: Reinhold Publishing Corp., 1962); and *A Wanderer in the City* by Walter Schwagenscheidt.

Second Stage: One Family and Its Interaction with Its Environment

This stage of the research must be conducted *after* the form determinants have been investigated, so that the interviewer can understand a family's narratives of daily activities without interrupting the discussion. For example, in the research on Vastrapur, when the mother, Sara, described "going for sticks to use in cooking," the interviewer knew precisely where she had walked without referring back to the base map.

A clear format for gathering this research has been developed so that styles of life discussed in various case reports might later be compared. The second stage begins with a detailed survey of the family's physical possessions. This is followed by a study of their home in relation to their daily life and to the surrounding environment. Interviews follow in which all movements of the family for one week within the neighborhood are plotted on an overlay to the base map. Finally, an exact description of the family's activities for one day is recorded. These last two interviews constitute the principal anthropological basis of each case study, and their importance is primary in formulating conclusions.[4]

4. Herbert Gans, in *The Levittowners* (New York: Pantheon, 1967) offers an excellent sociological approach to this problem. It is important to note that whereas Gans's approach is essentially personal historical narrative (without illustrations), the anthrophysical approach emphasizes visual form.

Third Stage: Conclusions and Reflections on Future Housing Form

Assuming the interviewer possesses some training in architecture or planning, it should be possible at this stage to make limited judgments about the appropriateness of the physical form to a family's (and to society's) needs. Beyond this, a few suggestions for future study of housing environment should emerge.

This general three-stage approach has been specifically applied in this study to two case reports, one on Parkview Place, a private-street neighborhood in St. Louis, and one on Vastrapur, a village outside the suburbs of Ahmedabad, India. The selection of these two housing areas, each located in a distinctly different culture, was not accidental. It was felt that the anthrophysical approach could best be tested by applying it against widely contrasting environments. Furthermore, the conclusions would be the more noteworthy because the neighborhood forms themselves represent strong alternative solutions to the general problem of ground-level housing for families with children. Naturally the conclusions reflect broad sociocultural differences, and while this limits direct comparison, it does not prevent observations concerning how the physical form is an outgrowth of the societal need.[5]

The methodology is now in the process of being "tightened," and work is underway to apply the approach to new subdivision development. A future study, comprising four case reports, will be a comprehensive illustration of the approach that the present book serves to introduce.

The serious researcher will not expect the anthrophysical approach to provide shortcuts for design or to produce startling new discoveries in urban form. Since the methodology is grounded in a humanistic outlook, its discoveries should be related to man's old but proper quest to understand, order, and then enjoy his physical environment.

5. A third case study, of Ichinomiya, a new workers' town in Japan, has not been included in this book.

1. *Entry Gate to Parkview Place*. The gate symbolizes where the city ends and the private street begins. At this back entry to Parkview Place, signs reinforce the message.

II   The Private Street System of St. Louis

EVERY city has a certain physical form, which helps to give it a flavor or quality distinct from other urban centers. In this respect, the private streets or places of St. Louis are worth a special and serious study, for they are the most impor-

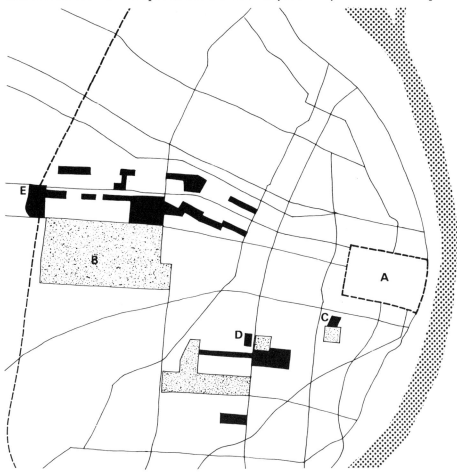

2. *Private Street Neighborhoods.* This diagram map of St. Louis locates the principal private streets of the city, all built before 1905. (A) Downtown core; (B) Forest Park (other parks are similarly shaded); (C) Benton Place, the first private street, built in 1867; (D) Shaw Place, the first street utilizing a wide park strip; (E) Parkview Place.

tant single physical development in St. Louis's architectural fabric, and they constitute a unique contribution to concepts of urban housing.

Most simply defined, the private street is an area in which the street (or street system) is owned by the residents rather than by the city, and it is distinguished by two important characteristics, one physical and one legal. The physical characteristic is the closing or blocking of the street at one end, which prevents through traffic and gains privacy for the inhabitants. Under normal city legal policy, closing the street to public traffic means that it must be privately maintained by the residents rather than by the city. This maintenance is insured through the use of the distinguishing legal characteristic, deed restrictions—written clauses bound into and made a part of the legal description of the property (and thus binding on all future sales of the property). Restrictions on the deeds of properties are not, of course, an invention of the private places. Even as early as 1847 the following restrictions appeared on deeds within a new subdivision in St. Louis: "No butchery, tallow chandlery, soap factory, steam factory, tannery, nine-pin alley, or any other offensive business or occupation may be set up or carried on within this property." One year later, another new housing addition, besides having all of the above restrictions, explicitly defined offensive business as "bawdy houses and distilleries."

But the private street uses deed restrictions in a new way. Within the private place every owner is required by his deed to belong to an association consisting of all the land owners. This association is empowered to make regular assessments for the common maintenance of trees, parks, and gates, and the paying of night watchmen. The deed restriction also makes it possible for the association to levy special assessments for street repairs, lighting, and sewers. More importantly, the restrictions limit use of the property to the single-family dwelling unit.

The differences between these deed restrictions and the use of zoning laws are obvious. An area zoned for single-family residency by a city council can easily be rezoned when the city council so desires (and even the best city government is not immune from political pressures). A deed restriction, however, is a lien against the property, and without the consent of every property owner within the association, the restriction may not be lifted. In addition, the lien takes precedence over other debts. For example, if the property is sold for reasons of bankruptcy, the unpaid assessments are deducted first from the proceeds.

Both the deed restriction and the physical act of closing the street have important architectural consequences. The most significant of these is the gate, a link where the private street meets the rest of the city. By the use of iron gates or simple chains, the streets may be alternately closed or opened at either end. At one time, every one of the principal private places in St. Louis had a gate. Other formal ramifications include: 1) a standard setback of the houses, 2) restriction to one common building material (usually brick), 3) the use of park strips in the middle of the street and the careful preservation of trees within this common area, and finally 4) the use of alleys for all services—garbage collection, telephone and electric lines, garages, and so forth. All of these form requirements can be (and have been) written into property requirements by using the deed restriction.

In capsule, the private place is an urban street restricted to single-family dwellings, closed to through traffic, and maintained by the use of deed restrictions. The place is characterized by its quiet, parklike atmosphere, the unity of construction within the area, and the gates which serve as a link between the place and the rest of the city.

The natural question of why the private places were built in the first place requires an understanding of St. Louis in the 1850s. Mr. George Brooks, former director of the Missouri Historical Society, has formulated an interesting hypothesis. He states that the rapid growth of the city immediately before and after the Civil War made it virtually impossible to construct a fine residence anywhere near the downtown core. The fantastic explosion of industry, commerce, and population simply overran any efforts at fine residential construction. It was not that one's property values did not go up—it was the bother of it all! No sooner would you get the furniture in the new house than you'd have to move and start all over again! Lucas Place (which was not a private place) was such a development; today only the Campbell House reminds us of the elegance of this street in 1850. Mr. Brooks suggests that the private street was a natural outgrowth of the desire for peace and quiet in an upper-class neighborhood, for freedom from smoke, industry, and noisy commercial establishments (and distilleries and bawdy houses).

Three other factors appear at least partly responsible for the private street concept. One is its natural appeal to a desire for status. To the residents of a private street it is quite clear they are in and everybody else is out. A poster of the late 1890s advertises Westminster Place by stating: "Entrance Gates are now being erected at Boyle & Taylor Avenues, thus giving the entire Place an air of privacy." The second factor is convenience to work, shopping, and schools. In a present era of rapid transit we forget the slow pace of the horse and buggy (although the highway at 8:00 A.M. is no joy).

The final factor is an elusive one that might be called "civic pride." The orig-

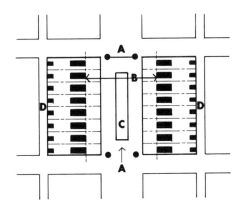

3. *Features of Private Streets.* (A) Gates, sometimes alternately closed and opened at each end; (B) common setbacks for all houses; (C) park strip maintained through yearly assessments on all residents; (D) alleys, used for all services, including telephone lines.

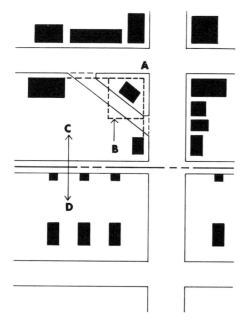

4. *Zone Breaking.* At critical intersection (A), an office building (B) now stands. Proposals called for removing this two-storied building and erecting a drive-in "Jack-in-the-Box" restaurant with angled entry. The clear demarcation between commercial area (C) and a private street neighborhood (D) would have been destroyed, and automobile lights would soon have affected all three houses shown. These plans were blocked, and the neighborhood's integrity was saved.

inal owners of houses on private streets also lived within the city because they wanted to. As well-to-do homeowners, they could have moved farther out, but again and again references to civic and cultural events crop up in old accounts of life within the gates. The private streets were showplaces in the city; guests such as President Cleveland on his visit to the 1904 World's Fair not only toured the private places, but stayed overnight in several of their houses.

The years of their development, from 1867 to 1904, mark the private places as clear forerunners of our present cul-de-sac streets, but to claim a direct relationship would be misleading. The *closed street* is a better term for the private places, and while it is true that their growth was fostered by developers, they were never subdivisions on the scale that has become common since World War II.

Whatever the reasons for developing the private street systems, the fact is that they are the unique conception of one remarkable man, Major Julius Pitzman. A surveyor, he laid out all forty-seven of the private streets within St. Louis. At twenty-two Pitzman was the owner of the finest surveying firm in the west, at twenty-seven a major in the Union Army. Seriously wounded in the leg in the Civil War at twenty-eight, he became surveyor for the city of St. Louis. So complete were his private surveying files that they were copied *in toto* to become the first official files for the city. *The Encyclopedia of the History of St. Louis*, published in 1899, states that "Pitzman's Records are known to all property owners. By means of these 'Records,' the new can be fitted to the old, as the key is to the lock. His office is probably the only in the U.S. where the present lines can be made to connect with the old lines, as they stood when Louisiana belonged to France."

Pitzman was thirty when he designed and executed the first private street in 1867—Benton Place, a small street on the north side of Lafayette Park in south St. Louis. Unfortunately, he did not record how he evolved the concept. It is known that he spent a year in London recuperating from his war injury. A surveyor by profession, he undoubtedly studied the English private squares. Surely he was aware of the need for fine new residential areas within St. Louis —he later built his own house in Compton Heights. Perhaps he was influenced by his friend, Henry Shaw, St. Louis's wealthy Englishman who could never quite believe things were better in the new country. Perhaps too, the owner of the land on which Benton Place was built, Montgomery Blair, was influential. More likely, Pitzman combined the idea of the English private square with that of deed restrictions that could insure privacy and quiet, and came up with the concept of the private street on his own. In any case, with the layout of Benton Place, Pitzman's reputation as an urban planner of first rank was assured. At first glance Benton Place is not a striking departure from other residential streets of its day. But a closer look shows all of the necessary ingredients: privacy obtained by permanently closing one end of the street, a private common park strip in the middle of the street (maintained by deed restrictions), and mansard-roofed French town houses built at a standard setback distance from the street. Today Benton Place has fallen from grace. Its gates are crooked and its gravelly street pocked with holes, but if it is now difficult to understand the importance of Benton Place, such was not the case in 1867, for other private streets sprang up immediately.

One of the most interesting of these is Shaw Place. Here the southern gate is always kept open and the northern gate kept closed. The most unusual feature of Shaw Place is the size of its central park strip; over one hundred feet wide, it is a natural playground for children. Developed by Henry Shaw, the street was never pretentious. Its houses are simple Victorian structures raised four to five feet off the ground and often containing beer cellars in the basements. The

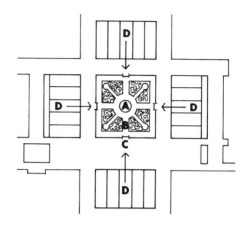

5. *English Residential Square.* A predecessor of the private street concept, and undoubtedly seen by Pitzman during his 1865 visit to London. (A) Park for residents of square, (B) paths, (C) gates, (D) houses.

6. *Benton Place, 1867.* The first of the private places in America (Drawing from *Dry's Pictorial Atlas* of 1873). Note upper-middle-class townhouses in French style.

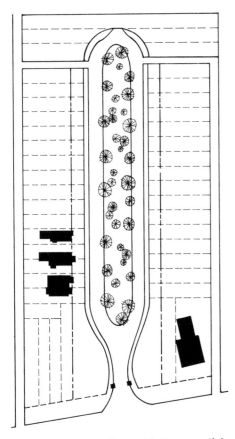

7. *Plan of Benton Place.* All the essential features of the private place are evident in this plan from 1868: a common setback from the street, a park strip, service alleys, a gate system (the north end connection of the street was never built).

houses have one feature of strange interest—Shaw bricked in one window in each house. The reason is simple. Formerly in England, country estates were taxed annually on the number of windows each house contained. One week before the assessor arrived, it was common practice to board up the windows and thus reduce the tax, and in honor of this custom, Shaw had one false window placed in each house, and then permanently bricked in.

The grandest of all private streets was Vandeventer Place, begun in 1870, only four years after Benton Place. A contemporary called Vandeventer Place "the cradle of our proudest and most substantial gentry." Developed in open farm land by three civic leaders led by Charles H. Peck, the place was too blocks long with a park in the middle. Everything about it was grander than anything before or since in St. Louis history; even the deed restrictions were demandingly stringent. Controls confined permissible land uses to single-family residences, explicitly excluding not only the customary tanneries and breweries, but public museums and schools as well. Any deed change required the unanimous consent of the Vandeventer Place Association, to which all the property owners of course belonged and which, among other things, had powers to enforce the scrubbing of steps twice a week and the hanging of three sets of curtains at each front window.

The original Vandeventer Place homes were soon outstripped in elegance. The fabled H. Clay Pierce mansion cost an estimated $800,000 to build (in today's money, approximately $3.5 million). Of course such mansions were built by the city's finest architects, and Vandeventer Place even had one home, the Lionberger House, by the most prestigious American architect of his age, H. H. Richardson. All the houses on Vandeventer Place had services in the rear—stables and groomsmen's quarters opening off the side streets. Pierce's home (and those that followed it) went further, with a separate house on adjoining Enright Street for twenty-two servants and a kitchen detached from the main house but connected by a heated underground tunnel. And of course Tiffany glass was used throughout.

The eighty-six residents of Vandeventer Place were the social elite of the city. Besides the Pierces, residents included the Catlin and Millikan families and Missouri Governor David R. Francis, president of the World's Fair and later ambassador to Russia. One of the more interesting accounts of life within Vandeventer Place is contained in the March 1925 issue of *The Censor*, a pleasant magazine of social gossip. Mrs. William B. Kinealy recalls her youth in Vandeventer Place:

From childhood I remember two lasting characters of old Vandeventer Place. One, old Mike, the private watchman, who bought whistles for the little girls and clubs for the boys; and Jake Scholtz, the head gardener. It was a treat to visit him in his cottage and have him tell us of the last big snake he killed right where the Walker home later stood.

The Watson Farr home had a secret closet in the window we children looked upon with awe. We were told you had to knock and it would fly open. The Newman house had those deep windows on the side recalling many an 'I spy' game and also some experiences. One, the times we used to watch the Chinaman cook make biscuits (strange all the Vandeventer Place homes had basement kitchens) and the other, the night we hid Mike's ladder, as he was preparing to light the park lights.

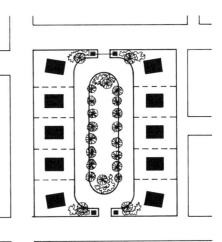

8. *Plan of Shaw Place.* Noted for its wide park strip, this private street is modest in scale. It was probably the first of the places to utilize two gates (the north gate is always kept closed with a chain).

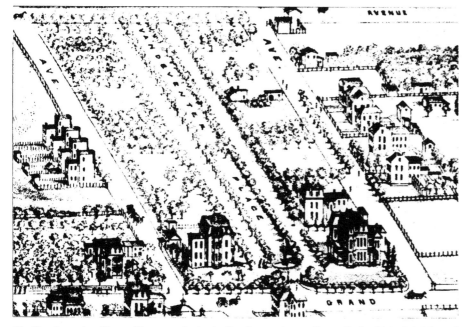

10. *Vandeventer Place.* The grandest of all private places (from *Dry's Pictorial Atlas* of 1873).

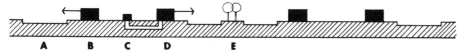

11. *Section through Vandeventer Place.* (A) Side street, used in place of alley for services; (B) servants' house, facing side street; (C) kitchen, with underground service to main house (D); (E) park strip, two blocks long, with large fountain at principal entry.

9. *Shaw Place Window.* In each house, Henry Shaw had one window permanently bricked in.

What St. Louisans in general thought of the private streets is best described in an unusual full-page feature in the *St. Louis Republic* of May 5, 1895, "Private Residence Places: St. Louis Rich in Paradisiacal Retreats." The article opens thus: "Everyone is surprised and delighted at our private places. . . . I recollect about three years ago a large delegation of journalists stopped over for a day in St. Louis. . . . It was the unanimous opinion of these gentlemen, who were naturally careful observers, that if St. Louis has not got a patent on the private-place idea, it is the only city which has developed it in a practicable, useful manner." The article goes on to describe Vandeventer Place—"As every model private place should be, it is a little world in itself"—and then Westmoreland and Portland places: "In the good days to come St. Louis will doubtless be a smokeless city. . . . But the owners in Westmoreland and Portland Places have solved the problem without waiting for the aid of science and municipal pressure . . . no bituminous coal may be used in this new area. And there is no smoke blown into these streets since they both are north of Forest Park and the prevailing breeze is from the south."

The streets of Compton Heights, it is noted, are "a delightful exception to

13. *Entry Gate Detail.* As entry gates became more prominent, elaborate and fanciful detail emerged, such as this eagle perched atop an Ionic column.

12. *Ironwork Traceries.* This wall encloses a house in Lennox Place and is typical of the ironwork found in St. Louis on the turn-of-the-century houses.

the American idea of parallel and rectangular streets. In London very many of the best residence thoroughfares are built in the form of a crescent. The object of this is to secure privacy, because through traffic naturally seeks a shorter and more direct route. This is, really, of course, the Private Place idea in another shape, and hence one may very safely include Compton Heights among the Private Places of St. Louis." The most interesting comment, though, is the following: "To the impartial observer at the present time it appears as though the only mistake made by the projectors of Vandeventer Place was the overlooking or rather underestimating of the rapid growth of the city. . . . Vandeventer Place is now very largely hemmed in by street railroads and business houses, and in the course of a few years this is liable to prove quite a detriment."

It proved quite a detriment indeed. By 1922 Vandeventer Place had fallen from the ranks of gentility, and the battle to destroy the single residence clause

15. *Brick Details in Vandeventer Place.* By 1880 brick masonry was a brilliantly developed craft, as evidenced by this house detail utilizing arches, sloped supports, terra-cotta friezework, and cut stone screen infills.

14. *Entry Gate Detail.* Ironwork applied to an entry gate.

16. *Kingsbury Place Gate.* A "big gate" of the 1880s, with a double entry and two columns flanking a well-known St. Louis landmark, the "nude virgin" statue (reputedly the daughter of a Kingsbury Place homeowner).

17. *The Lionberger House.* A home in Vandeventer Place designed by H. H. Richardson, America's best-known architect of the 1880s.

was on. In that year John and Sarah Harper opened a boarding house, and shocked residents reported seeing men lounging in shirt sleeves on the Harper porch. Even worse, some of the men were without collars and had the audacity to roll up their sleeves. The Residence House was closed, but the battle was eventually lost in 1947, when the Veterans Administration won permission to build the ugly hospital that now stands on the site of Charles Peck's home.

During the 1880s the private places went through a subtle change in which the gate increasingly became a symbol of status, and Kingsbury Place and Washington Terrace represent the apogee of the gate cult. The Washington Terrace gate is a medieval fantasy of red brick, complete with a gatekeeper's house, while the Kingsbury Place gate is Victorian classicism at its wildest. Ornate columns compete with griffins and urns, and in the center of it all the locally famous "nude virgin"—claimed to have been a daughter of one of the landowners.

The "snob appeal" of these areas, secure behind their gates, was fully exploited by the land developers. An 1890 advertisement for Kingsbury Place

18. *Washington Terrace Gate.* Adjacent to Kingsbury Place, this private street boasts the largest of all gates, complete with a gatekeeper's residence inside.

19. *Lewis Place Gate.* The end of the gate era is represented by the Lewis Place gate. Through the arch lies a genteel, all-black neighborhood.

20. *Park Strip.* By 1880 the park strip was not intended to serve as a playground, but rather as a showplace for the private place.

states: "The height above the low grounds is so great that the South and Southwest breezes come to it directly from the hills, cooled and freshened by their passage through the abundant foliage, free from miasmic vapors." By 1900, the era of the private street had reached its peak. The park strips were fully developed as great green swaths of elaborate planting, and the gates ranged from the robustness of Washington Terrace to the triumphal arch of Lewis Place. Even the alleys had developed a special character; high walls and planting combined to develop private back yards. Many of the fences and walls were beautifully worked compositions in wrought iron, and the houses themselves frequently had twenty to thirty rooms.

In this same year, a remarkable little book was published privately by an architect, William Albert Swasey. Called simply, *Examples of Architectural Work*, it contains numerous illustrations of his private street residence designs. The most interesting of these is the Lawrence house of No. 1 Westmoreland

21. *Entry Hall, Lawrence Home.* The landing of the central stairway in this grandiose home.

Place (torn down in the 1930s to reduce tax payments). Swasey describes the house:

> Originally built for Mrs. Siegrist, the owner's daughter, it is full of her ideas and requirements. The design selected was for a three-story colonial house, introducing two bay windows and a low third story, with flat tile roof and balustrade, *which the style called for* [my italics]. A tile roof, however, was finally adopted to please my client, and four feet added to the height of the third story to provide for a ballroom not originally contemplated.
>
> The interior of the house is most elaborate, a different style being required for each room, and no expense spared in their execution. . . . Some of the rooms are an Elizabethan dining room, an Empire library, and a Moorish smoking room. This lavishness of decoration and appropriate furnishing make it an interior hardly equalled in the West.

The Lawrence house represents a flamboyant eclectic style of architecture found in many of the private place residences. This house type is usually characterized by extravagance, great size, a pretentious plan with large entrance hallway, a third floor ballroom, and an exotic exterior treatment. As each new mercantile baron outdid his neighbor, eclecticism ranged from German castles and replicas of Greek temples to the Bixby's French chateau in Portland Place.

Two other principal house types may be found in the private places, and both are more stolid than their eclectic neighbors. One is the side-entry house. This type places its entry at the side of the house, allowing the living room to extend across the entire front of the residence. There are a great number of these houses in St. Louis, but perhaps the finest single concentration is on Lennox Place, which is filled with "Italianate" town houses. The other principal house

22. *Bixby Mansion.* Another flamboyant house, built as a showpiece.

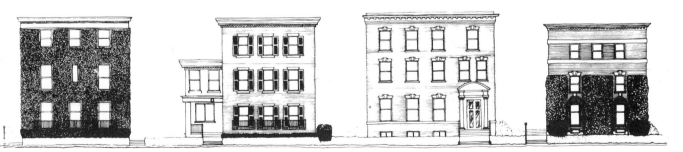

23. *Lennox Place, Partial Elevation.* The North side of Lennox Place probably constitutes the finest assemblage of houses in St. Louis.

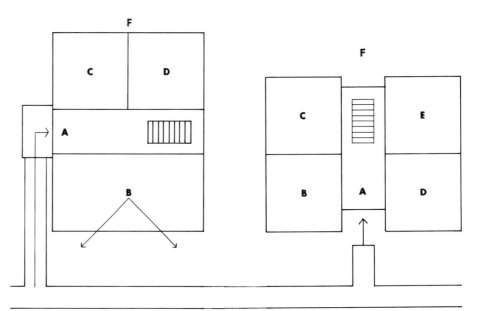

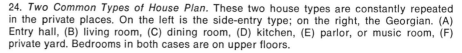

24. *Two Common Types of House Plan.* These two house types are constantly repeated in the private places. On the left is the side-entry type; on the right, the Georgian. (A) Entry hall, (B) living room, (C) dining room, (D) kitchen, (E) parlor, or music room, (F) private yard. Bedrooms in both cases are on upper floors.

type is the Georgian adaptation, with its straightforward, central-hallway floor plan. This type, of course, has shown great durability, for it is still being used today as a basic plan by subdivision developers. Architecturally its most interesting feature is an often beautiful entry porch.

In some respects the social prominence of the private place idea may be traced through these house-type categories. Vandeventer, Portland, and Westmoreland places all appear flamboyant, eclectic, and rich. Kingsbury Place is already more subdued, and the last of the great private places—Parkview, laid out in 1904, is quietly upper middle-class with a great preponderance of side-entry and Georgian house plans.

Of course, the private street concept has continued to influence residential development in the outlying St. Louis County area. There is even one suburban community which prides itself on having only private streets, with minimum public rights-of-way. But for the most part, developments since 1914 are private streets in name only. Largely suburban rather than urban in character, they represent not the concentrated townhouse forms of their predecessors, but rambling English garden estates, with one- to three-acre lot sizes, no sidewalks, and winding drives. When a forest already exists there is no need for privacy, and when the houses for several miles around are isolated on large land tracts, there is no need to establish elaborate boundaries with gates.

Meanwhile, although the private places within the city of St. Louis remain valuable elements of the city's architectural fabric, their future is unfortunately not encouraging. Most serious is the question of whether, in the path of growing slums, the private streets can maintain their quality. Remarks at the conclusion of this report will deal with this question directly.

26. *Washington Terrace Gate.* As it was when first built in 1880 (see also illustration 19). The gate as a developer's selling point is clearly seen in this vintage photograph.

25. *A Vanished Detail.* The ironwork gaslight of Parkview Place, later electrified, and now replaced entirely.

Parkview Place: An Urbane Private Street Neighborhood in Detail.

1. This case report, begun in 1963 and concluded in 1970, formed the pilot methodology for later studies.

Parkview Place[1] is the last of the great private street systems laid out by Julius Pitzman, and it is also the largest, comprising six streets, with 253 houses. Although layout and planning began in 1903, the majority of development work was done in 1904 and 1905 (in time to be seen by visitors to the World's Fair grounds, only one block away). From the beginning, Parkview Place was not so pretentious a development as its immediate predecessors. Lot sizes varied from 50 to 100 feet of frontage, with fairly uniform depth of approximately 150 feet. These smaller lots precluded obvious snob appeal, and the location, immediately adjacent to the newly built Washington University campus, encouraged professorial and professional occupancy. The great majority of the houses were constructed between 1910 and 1925, and by 1940 only seven lots were left undeveloped.

In January, 1970, the neighborhood consisted of upper-middle-income families, and in most of them the husband was professionally employed. The last annual meeting to elect new members of the Board of Trustees was attended by 188 persons (representing a substantial majority of homeowners). Interest was high in neighborhood schooling, police patrols, and a growing nearby slum, and discussion was spirited. This meeting and subsequent interviews with residents revealed certain social and cultural characteristics of the neighborhood. These were further substantiated by the "Parkview Survey" later conducted by two of the area's citizens, the results of which were published in 1970:[2]

2. "Parkview Profile" compiled by Robert Buck and D. Reid Ross, February 4, 1970 (mimeographed). Slightly more than 100 families (of 253) answered the survey. If nothing else, the survey is an example of the middle-class predilection for self-analysis: Are we moving upward? Yes.

3. This substantiates two conclusions drawn from personal interviews. A sizable proportion (about 20 percent) of Parkview families are either the original owners or grew up in the neighborhood. There is also a rising number of newer, larger families (with as many as thirteen children) especially attracted by the big but inexpensive houses.

The average family has 2.75 children in school from primary through graduate studies. 2.5 of these children are under the age of 18 but despite child orientation of the neighborhood, $\frac{1}{4}$ of the households contain adults over 65[3]. . . . 40% of the children attend public schools, 40% Catholic schools, and 20% other private schools. . . . The main reasons cited for buying houses in Parkview are that they

27. *Parkview Place.* This map shows all of the houses and streets within Parkview Place, as well as the roof forms and trees for the special quadrant in which the Steinman family lives. *Overlay 1* shows major features of the neighborhood: (A) entry gates (only those along the major arterial street to the east are kept open); (B) stone wall separating neighborhood from noisy street; (C) neighborhood parks; (D) alleys that carry all utility poles, services, garages, etc.; (E) pedestrian crosswalk, the boundary between St. Louis City and University City; (F) shopping street, Delmar Avenue (the intersection circled is shown in illustration 4); (G) the Steinman house; (H) A flat-roofed house, the only one in Parkview Place. *Overlay 2* shows all major trips to and from the Steinman house by the family during one week. The mother in her car made the most trips of any family member —forty-four in one week. The shaded houses are those of friends of the children. Note the large segment of Parkview Place that is never visited by the family.

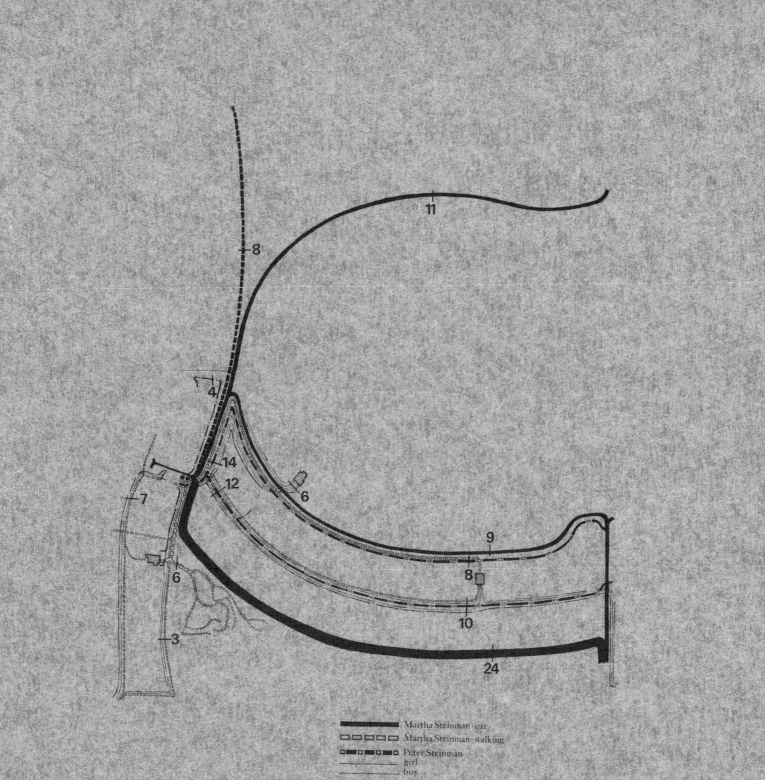

11

8

4

14

12

7

6

6

9

8

3

10

24

Martha Steinman—car
Martha Steinman—walking
Peter Steinman
girl
boy
maid

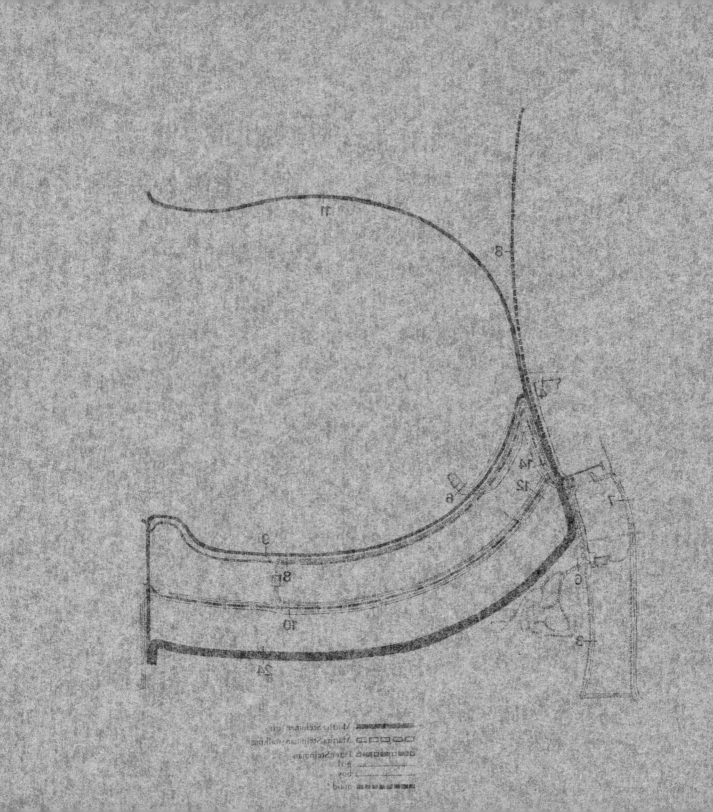

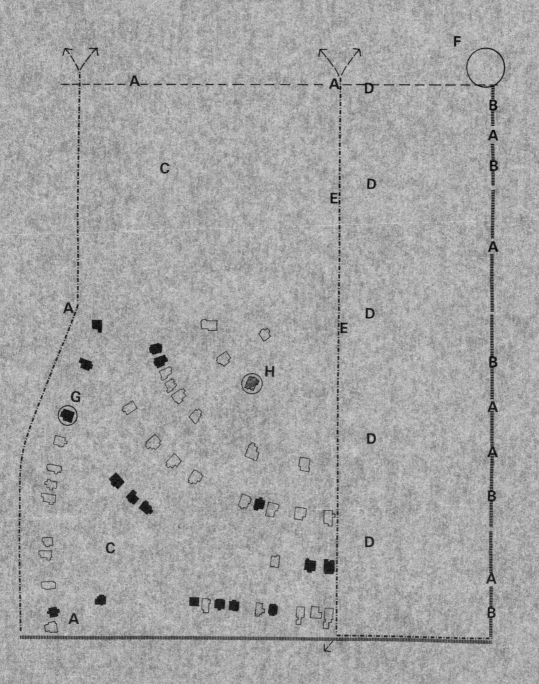

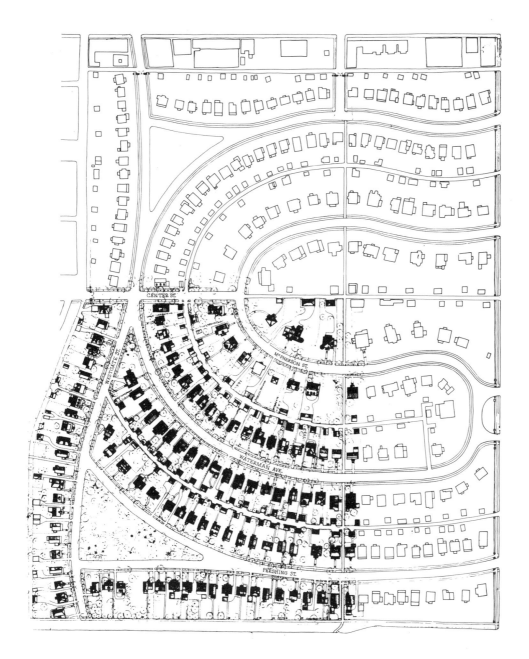

4. Thirty percent of the housing stock (82 houses out of 253) lies within St. Louis City; of these, a 1970 survey could not find one family that sent its children to the city public school. (An excellent Catholic school, St. Roch's, is within two blocks of the area.) On the other hand, the majority of the families within the University City section of Parkview Place sent their children to two public schools, each approximately a half mile away.

5. The issue of clean alleys reflects a peculiarly American tradition of home ownership that demands neatness in adjoining public property. By actual comparison with alleys in the nearby slum (ten blocks distant), the Parkview alleys are hygienic.

6. These conclusions are those of Buck and Ross. Liberally oriented families are hesitant to discuss racial questions in a somewhat public survey. In private conversations with homeowners I found considerable fear of racial change, not on social issues, but on grounds of loss of property value.

were large, of high quality, for sale at a good value for the price paid, and have good schools.[4] . . . The biggest problems with which the neighborhood is concerned are: (1) maintaining present school standards; (2) keeping dirty trashy alleys clean;[5] and (3) improving property protection. The problems with which the neighborhood is least concerned are: (1) the noise (which indicates it is a quiet neighborhood); (2) public services (indicating satisfaction with local government); and (3) fear of racial change (indicating modern social values).

In 1970 there was certainly an awareness in the neighborhood that the community was oriented toward the urban culture, and most homeowners took great pride in being residents of the city core. The survey stated that ". . . ⅔ of the families indicate they intend to remain five or more years in their present house, and only 20% intend to remain less than five years. Those that answered this question reflect little concern about racial or other change in the neighborhood or decline in property values."[6] Census data from 1960 (with some updating to 1970) indicates that houses range from twenty-five to fifty thousand

29. *Crosswalk.* The pedestrian walkway that bisects Parkview Place, this crosswalk allows access to shops at its northern terminus and to Washington University at its southern terminus.

28. *Alley in Parkview Place.* The brick paving is typical in St. Louis. The alley is maintained by the Parkview Place Association.

dollars in selling value, with two-thirds in the twenty-eight- to thirty-eight-thousand-dollar bracket.

A few comments are warranted on the area's deed restrictions, the principal means for holding the area together as a working social and neighborhood unit. Deed restrictions outlined in 1903 still govern the physical form. All houses must be either brick or stucco and must be set back a standard fifty feet from the street. Two-story construction is not mandatory, although seven thousand dollars must be allotted for construction, and from 1904 to 1925 this usually meant sufficient volume to necessitate two stories of building. As in all of the private streets, lot use is specifically restricted to single-family houses. Setbacks for sideyards are ten feet, although porte-cocheres and open porches may encroach slightly. The sidewalks are set back five feet from the curb, and this space is deeded for maintenance to the association, as are also two small common parks within the area. The arbitrary boundary line between St. Louis City and University City runs across the whole neighborhood. This line has been designated a pedestrian crosswalk, and is also owned by the association. Where alleys exist, all services, including garbage, incinerators, garages, and telephone and electric lines, are restricted to them.

30. *Private Park.* This photograph illustrates one of the two heavily wooded parks in Parkview Place. Note absence of play equipment.

7. Governor Caulfield was one of Park-view's outstanding citizens and wits. He frequently told residents that he would be glad to die so the community could increase the assessment, except that he did not relish the company of Satan. A lengthy interview with Governor Caulfield dates my own interest in deed restrictions and private streets.

The deed restrictions provided for three original trustees by name, with the unusual stipulation that no change or general increase in the annual assessments could occur as long as any one of these three trustees was alive. The last of the trustees, the former governor of Missouri William Caulfield, died in his ninety's in 1966.[7] The Board of Trustees immediately enlarged itself to include a member from each street or subsector as well as three members at large. Assessments also rose rapidly from an average of twenty-five to an average of seventy-five dollars per house per year by 1968. The trustees were charged with maintaining the easements, parks, alleys, walkways, streets, lighting, sewers, and police protection. Special assessments could be levied by agreement of a majority of the residents at the annual meeting.

Because of its size in area and numbers (estimated 1970 population was twelve hundred), Parkview more than any other private street area operates on a semitownship form of government, with good attendance at the annual meetings. As part of two large cities, the area has always enjoyed excellent relations with both municipal governments. Not the least reason for this is that every resident pays full taxes to one city or the other for water, sewer, streets, police, fire, and school services. The cities do supply fire protection and police protection when called; in general, however, the area still maintains privately a large proportion of its own services. Of those operated by the association trustees the largest current drain on funds is for police protection (more than fourteen

31. *Cross Section through Parkview Place.* (A) Street (public formal entry space); (B) sidewalk (semipublic); (C) raised yard (semiprivate); (D) house (private, interior); (E) fenced backyard (private, exterior); (F) garage (parking, garbage, services, semiprivate); (G) alley (semipublic, for neighbors only).

32. *Map of Special Study Area.* The base map shows house layouts with common set-backs, fences, garages, etc. *Overlay 1* shows all trees over six inches in diameter. The Parkview Place Association maintains and replants all trees along the street. *Overlay 2* illustrates the public-to-private hierarchy of space. Heavier shading represents more private areas (house interior), while lighter shading represents more public areas (the streets). Note how fences create sharp demarcation of space whenever private space is threatened (for example, along the public crosswalk).

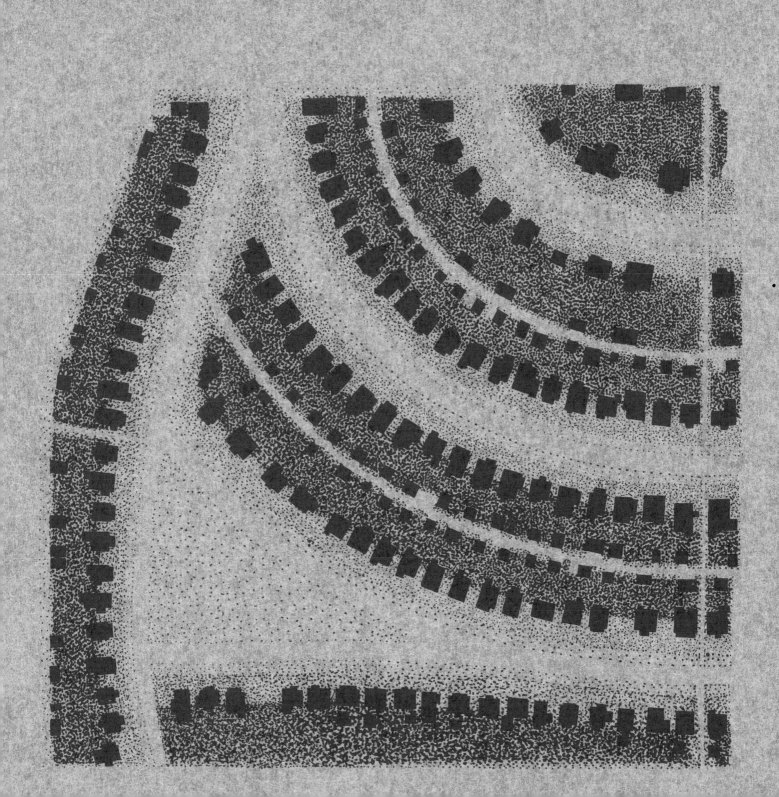

33. *Street Scene in Parkview.* A curving street, lined with trees, subdued and quiet in character—the exact image desired by the residents.

thousand dollars of a budget of twenty-two thousand). In addition, new street lighting was installed during 1969, for which a special assessment averaging seventy-five dollars per house was passed (to be paid in twenty-five dollar installments for three years.) Clearly, the area is viable and capable of accepting some change in structure. House values have enjoyed a small but steady increase in value, and residents are generally pleased with their environment.

Turning to the actual physical form of Parkview, it is readily apparent that the principal ordering elements come directly from the private street deed restrictions. The area has nine gates. The three to the north and south have never been opened. Those to the east off Skinker Avenue (one of the major north-south arteries of the city) have always been left open and are the general entries and exits for the neighborhood. The one gate to the west (on Center Street) is opened during the hours when cars and busses take children to and from school, but closed during rush traffic.[8] Thus, the common yet private ownership of the streets immediately sets a residential tone for the neighborhood.

The pedestrian crosswalk connects the area with Washington University on

8. When originally developed in 1905, the Center Street gate was always left open, as it had formerly been a farm-to-market road, and the trustees did not want to contest the case in court. Major Pitzman's ingenious reduction of Center Street to an alley, however, completely discouraged farmers, who soon bypassed the entire area with their produce wagons. In 1906 that gate was closed, and no question of legality was ever raised.

the south and with a shopping street, Delmar Avenue, on the north. The two common parks, edged with trees but somewhat open in the middle, form playing spaces for children but are kept clear of play structures—these exist only in back yards. The area thus has strong boundaries and great homogeneity.

The strongest visual aspect of the neighborhood is its trees. Because the association owns in common not only the streets, but also the five-foot planting strips between the streets and the sidewalks, a rigorous maintenance program can insure a band of shade along the sidewalks. The importance of the association's ownership and maintenance of trees within the public spaces cannot be overemphasized. Overlay 1 to illustration 32 demonstrates the form significance of this joint ownership, the trees in the public spaces far outnumbering those in the private. Since many of the originally planted trees (including all of the elms) are now dying after sixty years of growth, the trustees have initiated an extensive replanting schedule. Original plantings included sycamores, a particularly messy kind of tree, disliked by many residents—one resident annually requests at the homeowners' meeting that all sycamores be "cut down immediately." Since the streets were laid out in a curving pattern, the tree trunks form distinct and handsome columnar avenues along the walks. The curving street pattern accomplishes other form objectives as well: it prevents rows of houses, each with common setbacks, from becoming monotonous; it distinctly identifies the Parkview area as an entity within a larger urban pattern; and it further discourages speeding by motorists.

If Parkview's form has a critical flaw, it would be its insistence on almost total conformity—each house brick or stucco, usually three stories in height; each yard a segment of a fifty-foot band of green. There are no punctuation points, no gathering places, no landmarks. A series of sketches along the street vista would reveal little change from view to view. Only the alleys begin to have the sudden open vistas, the tight construction of fences and garages, and the unexpected punctuation points that characterize the village architecture of England. Not surprisingly, it is the alleys that most interest visiting architects looking for quaint charm.

As a group, the residents do not mind the seeming conformity of the neighborhood. Each year a community swimming pool in one of the two parks is urged by a few residents; each year it is roundly defeated at the homeowners' meeting. When a visitor objects to the difficulty of orientation by exclaiming, "I got lost—I couldn't find your house!" the residents calmly reply, "Yes, but you don't live here."[9]

9. J. S. Brunner in his excellent book, *On Knowing* (New York: Atheneum, 1965), p. 18, refers to "effective surprise" as the key to creativity. Parkview's "effective surprise" is *as a total experience*—that it exists at all within the larger framework of the city.

The Peter Steinmans: How One Family Responds to Parkview Place

10. The names have been changed.

The Peter Steinmans are typically Parkview residents.[10] Solidly upper middle class by almost any sociological or economic measure, both Peter and his wife, Martha, are college graduates, his education including a professional law degree. Peter's present income is in the forty-thousand-dollar bracket and includes stock investments and some inheritances as well as his lawyer's fees. As might be expected, the Steinmans are mildly conservative in political outlook, with strongly liberal views on integration and education. They voted for Nixon in 1968, but without much enthusiasm. They purchased their home in Parkview for thirty-two thousand dollars in 1958, when Peter was thirty-two, Martha thirty, and their two children, Agatha and Michael, were three and one. At the time of our interview, the children were eleven and nine, respectively.

Martha says, "Parkview represents what I liked best in St. Louis from my own childhood—quiet, trees, nice neighbors but not nosy ones, and a chance to enjoy growing up. Since we're both St. Louisans and both grew up in areas similar to this we always wanted to live in a private street." Both children attend public schools, but the Steinmans gave serious thought to private education before deciding on the University City school system (the nearest grade

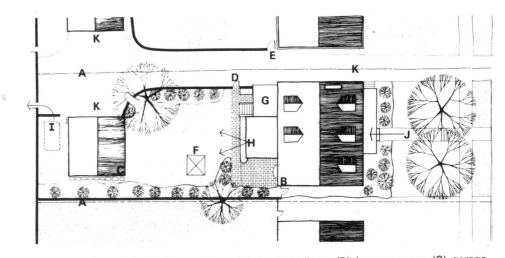

34. *Plot Plan of the Steinman Home.* (A) Property lines, (B) house proper, (C) garage, (D) gate built for visiting neighbors (E), (F) play equipment, (G) porch room, (H) kitchen window, (I) trash area, (J) entry sidewalk, (K) shared driveway (unusual in Parkview Place).

school is, however, heavily attended by children from professional families).

Despite the fact that they own two cars (a station wagon and a small sports car) and subscribe to *Time*, one should not infer that the Steinmans lead typically suburban lives. They are staunch supporters of the symphony, the art museum and the city zoo. Their library is extensive and includes the *New Republic* as well as legal journals. Peter and Martha buy fifteen to twenty hard-bound books a year—without belonging to a book club—and read regularly. In addition, Peter has served on the Board of Trustees for Parkview Place, and Martha is highly interested in local government and works for political causes within the inner city. They are proud of their "townhouse" life, and support Parkview Place verbally as well as financially.

Their immediate neighbors, Jerry and Mary Nichols, are close friends (Martha and Mary went to college together), and the fact that the two Nichols children are the same ages as the Steinmans' was an influence in the purchase of their home.

The children are together all the time, racing back and forth from house to house and running Mary and me ragged. Actually it's really convenient; she watches them when I go shopping and I watch them when she goes. And since we both use the same maid, Ginger, [who comes two days a week for the Steinmans and three days for the Nicholses], we can also sometimes go shopping together.

Ginger is a great help, but honestly I use her principally to watch the children. This gives me more time to cook for Peter. We like to eat late—after the children. It's when we have a chance to talk, have a drink, and enjoy each other's company.

The Steinmans entertain frequently—usually six to eight persons for dinner, or ten to fifteen for cocktails before a play or the symphony. Peter says, "We don't have to entertain that much. Actually my work [corporate law] doesn't demand the boss's wife be fed and all that. I think we do it because we like small groups with good conversation, and we like city life." "City-life" of course means friends, a generous house, good cuisine, safe environment and frequent enjoyment of St. Louis' cultural and sports events.

Their house clearly reflects their life-style. A typical Georgian variation, it contains nothing unusual as to plan, but it is remarkably suited to their needs. Off the upper stair landing Peter has a study which during the day becomes Martha's sewing room (and the guest room when guests come—infrequently). The bedrooms are large enough that each child has space for his own desk and playthings. The lack of a bathroom on the first floor disturbs no one in the family, Peter says. "The multibath mania is fortunately not one of our problems. In

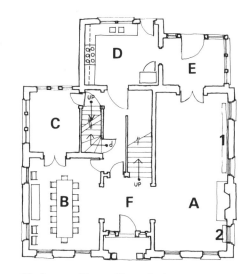

35. *Lower Floor Plan, Steinman Home.* (A) Living room, (1) books, (2) hi-fi equipment, (B) dining room, (C) pantry-porch, television, (D) kitchen, (E) kitchen-porch, (F) entry hall. Note also the second (servant's) stair, no longer used.

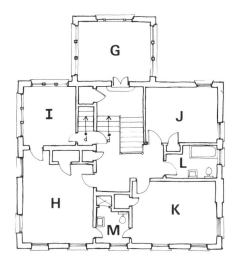

36. *Upper Floor Plan, Steinman Home.* (G) Mid-level studio; (H) master bedroom, (I) baby room, (J, K) bedrooms, (L) bath, (M) master bath.

37. *Entry to Steinman Home.* A typical Georgian house detail.

the evening if we're giving a party and want the larger bath kept clean, we let the children use our private bath."

The large first floor living space contains many books in cases as well as a built-in high-fidelity phonograph system. Peter likes classical music and listens to the phonograph at least twice weekly. The television is in the living room during the winter and on the screened porch in the summer, but they seldom watch it. Martha does enjoy "The French Chef" on the local educational channel, and Peter occasionally watches a sporting event. The set is largely for the children, although Martha discourages their watching it more than an hour a day, ("I'll slip on Saturday mornings sometimes"). The house has central air-conditioning, but they rarely use it because the screened porch is well shaded, and they both enjoy the yard.

The full basement is a feature they both like. Martha says, "It's amazing what goes down there. Mother's old set of china, Pete's wine racks, trunks and suitcases, the laundry, a freezer, a rake and hoes and all that stuff, paint brushes, screens, the furnace. . . . Best of all, it's dry and pretty clean and when the children were younger they used to play down there a lot."

It is on a social level that the house "works perfectly for us. The dining room seats twelve persons comfortably—without being too large, so that when we eat there at night it's still a nice room." The size of the living room allows cocktails for twenty people "with just a little crowding, which helps everyone talk better."

The kitchen was remodeled by the Steinmans in 1958. New cabinets, stove, ovens, and refrigerator were added. No architect was used, since Martha knew "pretty much what I wanted, and mother had this carpenter who had worked for the family before, so we called him in and he did the whole thing for us." In addition, the division by floors into one zone for sleeping and one for "living" suits the Steinmans, allowing the children to sleep without interruption by guests.

The children's special zone is obviously the back yard, where play equipment—swings and a sandbox—allow Agatha and Michael and their friends, Tom and Georgia Nichols, to play nine months of the year. With wooden fences surrounding their backyards, each family has privacy when desired. (The shared driveway is a somewhat unusual feature in the private-street neighborhoods—it serves as an identifiable physical tie between the Steinmans and the Nicholses.) When the Steinmans built a new fence after they purchased their property, they arranged the gates to facilitate communication with the Nicholses.

The Steinmans feel that their house, their yard, their friends, and their neighborhood—Parkview Place—suit their desires and needs perfectly. How then do they actually relate to this immediate environment? To understand this relationship, the paths of travel taken within Parkview Place by every member of the family (including the maid and the dog) were plotted for one week in June, 1965.

The pathway map (illus. 27, overlay 2) reveals several aspects of how the Steinmans in fact understand their neighborhood. The most unusual of these is that Peter walks through the alley to reach a downtown express bus stop in the morning, but he walks back from a different stop in the afternoon along the street. He says he does this because "It is a shorter walk to use the alley in the morning, and the street in the afternoon." Actually it is shorter in distance to use the alley both ways. The pathway map also reveals Martha's complete dependence on the car. In fact, there is a vast segment of Parkview that the Steinmans rarely see. It is thus clearly not part of their visual environment, but as Peter says, "We know it's there, of course, and we know what it looks like without having to drive through it. I guess you could say it's part of our psychological environment." This statement betrays both his educated perception and his acute awareness of property problems within the city. If any house within the area were to deteriorate, Peter would soon hear of it, drive by it, and urge action by the trustees.

A typical day for the Steinmans, (in the summer of 1965) begins at 7:15 A.M., when Martha arises, awakens the children, and begins breakfast. By 8:00, everyone is up and the family has breakfast together. Peter lingers over coffee to discuss the day's activities and their plans for Friday night, when they have been invited out for dinner. The children finish in a hurry, and because the sun is shining they are outside on the swings by 8:20. At 8:30 Peter begins his five-minute walk to the express bus, which takes twenty minutes to reach downtown. He usually arrives at his office between 9:00 and 9:30.

It is to be a casual day, warm (eighty degrees) but not too hot, with relatively low humidity (Martha hears this on the radio).[11] After feeding the dog on the back steps, Martha tells the children she will take them for a walk. By 9:15 she has stacked the dishes and dressed, and is ready to go out; but by this time Georgia Nichols has come out to play on the swings, and she and Agatha

38. *Walk Detail in Parkview Place.* The transition from public sidewalk to semi-private family sidewalk is given architectural treatment.

39. *Backyard in Parkview Place.* The fence on the right, adjacent to public crosswalk, gives complete visual privacy to this backyard.

11. In St. Louis the temperature frequently reaches ninety degrees and the humidity ninety percent in the summer.

40. *Backyard in Parkview Place.* Note the variety of treatment possible, as contrasted with illustration 39.

decide to play at the Nichols home while Martha, Michael, and the dog go for their walk. Michael wants to "see the University," so they walk onto the campus (Michael holds Simon with a leash). Michael identifies most of the buildings and is loquacious about his favorite, the Law School, "where Daddy went to school. The best place is the library. Even I could study in there because it's so quiet and smells good, too."

By 10:30 the walk is finished and Simon is penned in the back yard. Michael goes off to find Agatha, and Martha begins household chores (Ginger, the maid, doesn't come on Wednesdays). After cleaning and dusting the living room, Martha talks with her neighbor, Mary Nichols, on the telephone. Martha wants to go to a special grocery store (about a mile away) to buy fresh veal for that night. Mary agrees to watch the children, and Martha agrees to pick up some extra bread for Mary.

At 11:20 she dashes out to the station wagon. ("I'd be lost without this monster to drive. Sometimes I don't think I could find my way to Delmar Avenue if the car broke down.") At 11:55 she returns, and both her children along with Tom and Georgia Nichols parade in asking for lunch. "It's O.K. with Mrs. Nichols, we asked her." Martha is not elated by this news but she promptly fixes peanut butter and jelly sandwiches with celery sticks, milk, and a small dish of ice cream for each (including herself). At 12:40 she sends the Nichols children home (with Mary's loaf of bread) and tells her own children to play in their rooms. This is a fairly standard summer schedule, and both children have long ago learned that Martha likes a rest after lunch. Since their rooms contain books, paints, and record players, as well as the usual toys, they do not object to this "quiet time" (as Michael calls it) of the day.

After beginning preparations for her evening meal (veal parmigiano, with a fresh-baked loaf of Italian bread that she found at a bakery next to the meat market), Martha goes upstairs to read and to nap lightly.

She arises at 3:30 and tells the children they may play outside if they wish. As Martha is preparing the two separate evening meals (the children will eat a hamburger loaf), Agatha comes in, wanting to help cut up the vegetables. Michael is now playing in the park with several neighborhood children. Agatha and her mother decide to play some records and Martha lets Agatha choose. She plays a Beatles record, and although Martha is not fond of the selection (Peter bought the record mostly for parties), she doesn't voice a strong objection. Sounds of "Hard Day's Night" thunder through the house until Martha makes Agatha turn the record player down.

By now it is close to 5:15, and Martha is just ready to look for Michael when he comes in the door and both children race to the back porch to see the news on television. "Watching T.V. on a schedule" Martha says, "is still not the thing around here. If Agatha wants to see T.V. so does Michael and vice versa almost immediately. They'll see about fifteen minutes of the news and then decide suddenly to watch wrestling or whatever."

At 5:30 Mary Nichols telephones and suggests that Agatha come over for dinner right away because they've suddenly decided to take some of the neighborhood girls to see the movie *Heidi*. Agatha is delighted and so is Martha. "Mary's just like that. Always on the spur of the moment. Sometimes I dislike two extra kids for lunch at 11:55 but Mary never cares. I will say, she keeps both families hopping."

At 6:15 Peter comes home, and seeing Michael alone watching T.V., he sits down to talk with him. Michael's dinner is ready at 6:30, and they all gather in the kitchen and breakfast room while he eats. Martha continues working on her special meal (she won't tell Peter what it is). Meanwhile he sets the dining room table for two, feeds the dog, and then mixes a bourbon and water for Martha, drinks two scotch and waters himself, and keeps up a running conversation with Michael about the T.V. news on Vietnam. At 7:15 Michael goes out to play in the backyard, and Peter and Martha sit on the porch with their drinks, listening to music (in this case a Mozart concerto, although Peter was tempted to replay the Beatles album).

At 8:00 P.M. Michael is called in and told he should get ready for bed, but that if he wishes he can read for an hour in his room. Peter goes up with him to see that the instructions are started, and when he returns at 8:15, Martha has the meal on the table, including a bottle of wine.

Dinner is leisurely, and not until 9:20 do they finish their coffee. During this time the record player has stopped, and Parkview Place itself has become very quiet, with only an occasional car passing the house. They clear the table together and as Peter is jokingly asking Martha why she's gone so Italian lately (last week she served green noodles with clams), Agatha comes home from the movie, is promptly sent off to bed, and the house is quiet again.

Peter goes to his study to read, make out a quarterly tax statement, and prepare for a case tomorrow in court. At 10:00 he comes downstairs and together they watch the T.V. news, hoping to see a friend who is involved in a prominent local criminal trial as the defense lawyer. Disappointed, at 10:30 they have another cup of coffee as Martha finishes the dishes. By 11:00 P.M.

41. *Crosswalk.* The crosswalk has become a place for smaller children to play under supervision. Fences protect the private yard space from visual contact with pedestrians.

they are both in bed, reading, and by 12:15 their lights are out, and everyone in the Steinman family is asleep.

The above description of one day in the Steinman family's life reinforces and expands the earlier understanding of their life and their relation to the physical environment of Parkview. At this point, certain conclusions can be drawn:

1) The house plan itself works well for the Steinmans. During this typical day every room of the house was used, with the exception of the upstairs nursery (now used primarily for storage).

2) They are happy and pleased with the environment—life is relaxed, and one knows who he is in Parkview. Although their life is scheduled, it is capable of flexibility (Agatha's sudden trip to the movies; Martha's early decision to cook "something special" for Peter). Considering their maid, the neighbors' help with children, and the overabundance of young babysitters in Parkview, the elders' life is greatly freed for social activities outside the immediate neighborhood. These social activities are important to their life style.

3) An upper-middle-class family, the Steinmans find their house, its fenced backyard, the continuity of the streets, the use of the common park, the quiet at night, and the proximity to cultural advantages all amenities which they appreciate—even to the point of mild complacency.[12]

12. "Satisfaction with life" might be a better phrase than complacency. While issues such as Vietnam and the growing slum concern Peter and Martha, their primary concerns remain centered around their children and their lives together.

Peter sums up their lives, Parkview, and the entire idea of the private street system when he says, "It's quiet, it's nice, it's in the city, and we're protected from the slum. At least, I hope so—I don't think I could stand suburbia."

Conclusions: The Deed Restriction as a Planning Tool

Major Pitzman's design of Benton Place in 1867 was without question a brilliant innovation in urban residential planning. The legal protections afforded by deed restrictions, coupled with the closing of streets to through traffic, created pockets of quality housing within a haphazardly growing city. The private streets also fulfilled the social need for residences within the city suited to an emerging well-to-do mercantile class.

Although the private street's original attraction for developers lay in its obvious "snob appeal," the deed restriction is primarily a planning tool and not an end in itself. Deed restrictions have for many years been one of the principal legal means of preventing house sales to blacks. Like any legal proviso, however, a deed restriction is subject to court review. No blacks now live in Parkview Place—although the St. Louis Cardinals' pitcher, Bob Gibson, rented a house there during one baseball season. There is nothing in the deed restriction itself that prevents a sale to blacks. Lewis Place, in fact, is a private street composed entirely of blacks. However, only a small percentage of any minority group can afford to pay thirty thousand dollars for a house; those who can do so prefer to "protect" their investment (just as the middle-class white generally does) by buying into suburbia. A few private street areas have begun openly to encourage black ownership in hopes of stopping the growing slum.

Parkview Place rewards investigation because its size affords the best opportunity to study the private street on an appropriate scale for future urban planning. Moreover, it is decidedly upper middle-class rather than upper-class (the really wealthy can afford both a country estate and an apartment in a high

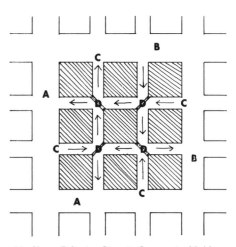

42. *New Private Street Concept.* Making streets one-way in an existing street grid by closing at points (D) prevents through traffic and creates a quieter residential area. This modification was successfully introduced in the St. Louis Rosedale-Skinker neighborhood.

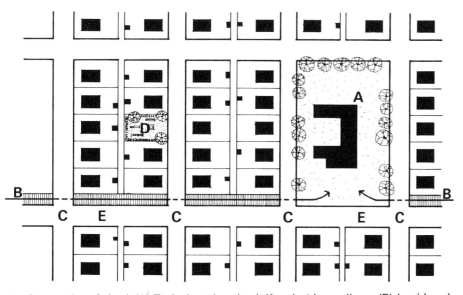

43. *Crosswalk to School.* (A) Typical grade school. If pedestrian walkway (B) is widened and improved for children, then the critical areas become the street crossings (C). If streets are closed at (C) then a private street area is created. The width of the park strip could create noise barrier for houses from busy street (E). Vest pocket park (D) could be maintained for total neighborhood through use of deed restrictions.

13. A corner site drive-in (illustration 4) was almost built at Parkview's northeast corner. Had this "Jack-in-the-Box" been built, it would have seriously impaired the cohesive form of Parkview.

rental tower). Parkview still succeeds as a residential unit for a variety of reasons. It is largely cohesive in form because its boundaries are clear, protected on the south by Washington University and Forest Park, on the west by another private street area, on the east by a major thoroughfare (Skinker Avenue), and on the north by a shopping street (Delmar Avenue).[13] In addition, its site adjacent to Washington University, coupled with its nonfashionable address, keeps its house prices well within range of professional salaries. Large enough to fight incursions successfully, Parkview Place has a private operating budget sufficient to initiate court action against those few property owners (usually absentee) who do not maintain their homes. More importantly, the size of the area's population gives it political leverage that aldermen cannot ignore. Parkview's most serious problems are its division between St. Louis City and University City, its lying in the path of a slum growing from the northeast, and its difficult and deteriorating public school situation. At the moment, though, Parkview Place is a successful, comfortable neighborhood in which to live. In the last analysis it works because those social forces which made it a speculative success in 1905 still operate today. Thus, the physical implications of deed restrictions *continue* to respond to the desires of its present population.

Beyond the consideration of Parkview Place as an area in which to live, what are the implications of deed restrictions as a tool for planning future city form? To begin with, deed restrictions are more potent legally than zoning ordinances. Former Washington University professor William Weismantel has suggested that the city might use deed restrictions as a public planning tool rather than a private one, and thus initiate neighborhood rehabilitation within the existing city. One obvious use for such planning would be to create stronger single-family residency requirements (the city's participation here would be in organization, free legal advice, and public meetings). Another would be the possible deeding of public streets to a private subneighborhood in order to stop through traffic and create pockets of housing with streets safe for children's play.

There is a third, seldom discussed potential for deed restrictions within existing urban residential neighborhoods, and that is in the development of walkways for school children, of park strips, and of vest-pocket parks, built on older streetcar rights-of-way or on cleared land. Land clearance authorities have never considered the deed restrictions as a means of gaining neighborhood support for a project. If the neighborhood association owns the vest-pocket

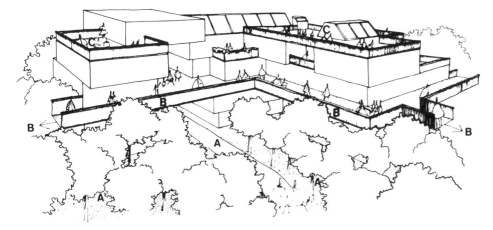

44. *Three-dimensional Planning.* Utilizing the deed restriction as a tool for implementation of planning ideas: (A) ground-level kept for public use; (B) mid-level (perhaps third floor) inter-connected from building to building; (C) roof levels devoted to play-schools, laundromats, etc.

park, the homeowners have a natural interest in maintaining its value for their children. The neighborhood association can cause each of its members to have a legal and binding interest in maintaining walkways or parks in an area, even those not in front of their homes. In illustration 43, if the major street (E) has a high noise level, the protection of property values along the school pathway (by planting setback, walls, etc.) will in the long run also protect a house within the block.[14]

Deed restrictions have an even greater potential in planning *new* urban environments. Whereas zoning is principally a two-dimensional tool, deed restrictions can be used to reserve and maintain air rights, underground utilities, and linkage systems. Indeed, their greatest potential is in three-dimensional multi-use planning, to enforce a mixture of commercial, public, and private uses both two-dimensionally on the ground and three-dimensionally through vertical restrictions. For example, deed restrictions could create streets-in-the-air that link a total neighborhood (as has been partially done at Park Hill, Sheffield, England). Buildings over eighty feet high could be required to devote roof space to public use (the Hong Kong rooftop schools show an instance of this). To date, restrictions in high-rise speculative construction have been centered around the car (so many spaces per office module). Little concern has ever been shown the worker himself. No city better exemplifies this than Clayton, a satellite of St. Louis. Accidental office cities such as Clayton

14. The famous "Radburn Plan" by Henry Wright and Clarence Stein shows one of the housing forms possible when a pedestrian walkway system is used on a large scale—Wright grew up and was educated in St. Louis, and was intimately aware of Pitzman's private street systems.

might well have written a "third-floor public space" clause into deed restrictions by which the speculative superblocks, interconnected at thirty feet above grade, would give shoppers, the workers, and children a chance against the automobile. The river walkway in San Antonio is an example of what might be preplanned below principal ground level through use of deed restrictions.

If deed restrictions are a feasible tool for improving environment, how could they be legally adopted within existing city-planning frameworks? This is a question beyond the scope of this report, but obvious incentives for voluntary private participation could be developed through tax reforms and tax relief. Neighborhoods might be induced to join binding associations to maintain streets, planting, lighting, and so forth, if given a reasonable tax reduction as encouragement. This principle applies equally well to store owners in a neighborhood who can join together to implement street lighting, planting, arcades, sign control, and off-street parking.

One conclusion is inescapable: the deed restriction concept and its present physical manifestation in the private street system of St. Louis deserve far greater study, not only by city planners, but also by anyone—sociologists, economists, and lawyers—interested in making the city once again habitable.

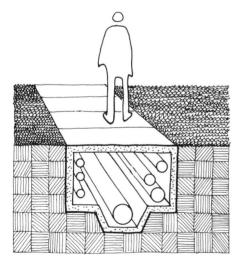

45. *Utility Right-of-Way.* A simple and straightforward possibility for use of the deed restriction.

III   Vastrapur, near Ahmedabad, India

Introduction: Village Life in India

VASTRAPUR is a small village of five hundred persons lying just outside the suburbs of India's urban textile city of Ahmedabad. Like many of its counterparts, Vastrapur casts a special charm on visiting western architects.

Dark verandahs filled with sleeping figures, homogeneous sun-whitened mud-brick housing, courtyards with one oak and black pools of shade, through which flow sari-draped women—such are the hazy, remembered visions of form-conscious architects. Nor are connoisseurs of folk arts and crafts immune to this nostalgic spell. Hand looms, potter's wheels, carved door lintels, festival days of brilliant color, frenzied dancing, and peppered tongues—all conjure up images of a simpler life where form followed social need and function. Such visions, recollected in tranquillity, are misleading. In abstract and pure form, villages such as Vastrapur are beautiful. But in the real world of a changing India, they no longer respond to the aspirations and new needs of the society. Resignation was an older way of life, but it was never shown to be a satisfying experience. New India demands more.

Vastrapur was chosen for this study for three reasons. First, as an older Indian village, with social patterns presumably reflected in its physical form, it provided a good test for the anthrophysical method. The method should illustrate the ways in which housing forms differ markedly from culture to culture, due to differences in life styles. Second, dealing with ground level housing, the conclusions reached should reveal similarities as well as contrasts with U.S. suburban form and should pose interesting questions for further consideration. And third, because of its proximity to Ahmedabad and the certainty of its being engulfed by that city, Vastrapur provided a timely situation for investigation by the students of the Architectural School in Ahmedabad. This made rapid, in-depth study possible and meant that the conclusions would have some bearing on the emerging social problems of India.[1]

To begin investigations into the social and physical format, it is necessary first to understand a few facts about India's agrarian life in general as well as about Ahmedabad and its geographic area.[2]

1. *India.* The village of Vastrapur is on the outskirts of Ahmedabad.

1. The data for this report was collected in October and November of 1964, during a six-week teaching stay at the School of Architecture in Ahmedabad. I am indebted to the third-year class of that year and to the school's director, B. V. Doshi, for their assistance.

2. Patrick Geddes's *Notes on India* (London: L. Humphries, 1919) is an excellent introduction to problems of village planning in India. More recent, and equally interesting, is Oscar Lewis's *Village Life in Northern India* (New York: Random House, 1958).

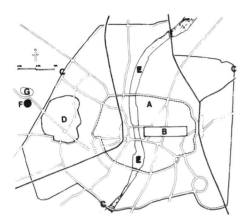

2. *Sketch Plan of Ahmedabad.* India's textile center, Ahmedabad is an old city filled with Moghul architectural treasures. (A) The old city, (B) the central business area, (C) present city boundaries, (D) The University area, (E) the Sabarmati River, (F) Vastrapur, (G) Vastrapur Lake.

3. A superb account of the architectural treasures of Gujerat is contained in Klaus Herdeg's *Formal Structure in Indian Architecture* (Ithaca, N.Y.: Cornell University Press, 1967), some of these being: Adalaj, a step-well; Kapadvanj, a village reservoir; and Surkhej, a religious complex.

Ahmedabad, a city of approximately one-half million persons, is three hundred miles north of Bombay, and is the largest city of the state of Gujerat. As in most large subareas within India, the state has its own language, Gujerati, as well as its own style of dress (involving draped trousers for men and women similar to what Americans normally think of as Punjabi attire). The educated class speaks English as well as Hindi, which is similar to Gujerati.

In climate, Ahmedabad is very hot in the summer (ninety to one hundred and ten degrees), and when it rains (during the monsoon season in late spring) it rains constantly. After the lighter fall rains, there are the two delightful months of November and December. January and February are cold, but rarely below freezing, and as in almost all Indian towns south of Delhi, the common man sleeps outside most of the year.

Ahmedabad itself has a number of giant textile mills, producing the majority of India's cotton fabrics. The present city is filled with Moslem influences, containing not only mosques, but also a number of other buildings with traceried windows of great beauty. The Moghul empire left many of its greatest architectural monuments in and around Ahmedabad.[3] Nonetheless, Hinduism is the principal religion, with its strong social caste system. The Jains are the highest caste within the system, and their power and wealth is staggering. Approximately seven intermarried families own all of Ahmedabad's textile mills. Two of these families—the Shodhan's and the Serabai's—have villas designed by Corbusier. (The Serabai "compound" contains a half-dozen houses, with outbuildings, guest houses, pools, servants' quarters, and so forth.) At the opposite end of the scale are the farmers, water carriers, potters, and general laborers.

Fortunately, even at the village level, the government has made real efforts to provide both for better health and for open educational opportunities. Thus, although there is no adult social communication between castes at the village level, there is normally only one school, which all children attend. The state government is also organized at the village level, each village containing a *panchayat* or central office with representatives, similar in structure to an American city's council and mayor. The *panchayat* differs from U.S. local governments principally in that it has a direct connection with the state and national administration. This means participation and interest within the village is increased (and concomitantly, political communication occurs between castes).

The most important aspect of all village life in the state of Gujerat is its agrarian nature. All villages are oriented to the land, and to outdoor life. More than three-quarters of all family activity occurs outdoors, in and around small public spaces.

Vastrapur: Its Physical Form and the Social System

The village of Vastrapur will soon be engulfed within the suburbs of Ahmedabad. Yet at the moment it retains its agrarian life and its indigenous form, created through one hundred years of slow growth, decay, and renewal. The color is homogeneous—houses are of mud-brick and mud-plaster, sun-bleached almost white, with roofing of rough, wooden poles and thatch. Small in scale, the houses (with only one exception) are one story in height and never more than two rooms in size, with a verandah attached. Of the villages's 119 houses all but 4 contain verandahs. Sameness and continuity are thus stressed: low cactus walls surround semiprivate storage yards, and one "street" looks much like another with patches of green grass intermingled with sandy dirt walkways. The village is loosely organized, and has obviously grown (or shrunk) as needed. A look at the general plan confirms this.

The village is connected to Ahmedabad by a dirt-gravel road which changes to asphalt about one-quarter mile to the east (roughly the present edge of the suburbs). This road, used by pedestrians, carts, bicycles, and intermittent busses, continues past the village to the lake, and beyond to neighboring villages. Vehicular traffic thus bypasses the village proper, and Vastrapur's streets are in reality pathways—those areas not enclosed by cacti—used by goats, dogs, cows, pedestrians, and bicycles.

It is clear that communal activity is dominant over private. This is illustrated by the scarcity of elements of form one normally associates with private space. There are few physical boundaries—one space becomes another through use and not by design. This is demonstrated not only by the circulation system, but also by the amorphous pattern of the housing. Likewise, the "public places" appear to have no formal site locations. The *panchayat* sits in a dusty open place beside the road; the school is in the center of the village, but it has no distinct "front entry"; the two wells (which everyone uses) are only roughly situated for convenience; and while the two temples are both beside the road, their location relates little to the village form.

Of the two important manifestations of natural form, the lake and the trees, one is conspicuous by its near absence. While at one time the Vastrapur area was heavily forested, there are today only a few trees within the village, and new growth is hampered by overpasturage, both by cows and goats. The lack of good wood also directly affects village construction. There are few doors or windows, and roofing is framed with crooked branches never more than eight to ten feet in length. The villagers themselves realize the need for more trees— for the benefits of shade, fuel, dust control, and lake maintenance. They see no way, however, to pay for tree planting.

Although at present the lake poses a mosquito problem (and although during the monsoon it rises ten feet, flooding a large area), its inherent physical, social and psychological values should not be overlooked. The lake is the original reason for the village's location, having always been the sole water source for the buffalo of the dairy farmers and for the one or more buffalo cows owned by 80 percent of the families for milking purposes. (Milk is the principal food of India's children, and even sacred animals may be used for this purpose.) With proper care, the lake could again become the village's major asset instead of the muddy and unclean place it now is. In his excellent *Notes on India* (1919) Patrick Geddes discusses tanks (or lakes): "The many values of the village tanks are seldom appreciated by sanitarians who consider them only as breeding places for the malarial mosquito. . . . Cleaning out these tanks is a serious annual expense . . . planting the banks with trees would greatly reduce this expense, as the tree roots hold the soil from being washed down by the rain . . . the cooling value of these reservoirs has an appreciable influence on health and comfort. . . . Much can be said for them during the long dry season in maintaining the water level both in wells and in the soil."[4]

It is only when one begins to consider the caste system and its social implications that a sense of organization begins to emerge in Vastrapur's physical form. There are three castes in the village, all of approximately equal importance in the hierarchal scale of the social system—dairy farmers (com-

4. To Geddes's proposal to plant trees at the lake's edge, we might add proper cleaning, lining, deepening at the edges, stocking with ducks and fish (which would eat the mosquito larvae), and the planting of algae and other green plant life within the lake. The predominant Indian solution for mosquito control, unfortunately, has been to rely on spraying.

3. *Base Map of Vastrapur.* This drawing was used in all studies of the village. The potters' courtyard (P) is the area of special study. The arrows represent verandahs and the direction in which they face. *Overlay 1:* The "green vegetation-to-sand" hierarchy. The solid areas represent trees; the dotted areas, grass. Note that grassy areas (used largely for water buffaloes) are defined by cactus walls. *Overlay 2:* The "private-to-public" hierarchy of space. The darker areas are more private; the lighter areas more public. Note that both overlays reinforce the contention that life in Vastrapur is public-oriented, the inhabitants generally walking everywhere except where cactus walls prevent their passage. Variety comes from the constantly changing scene in front of each verandah.

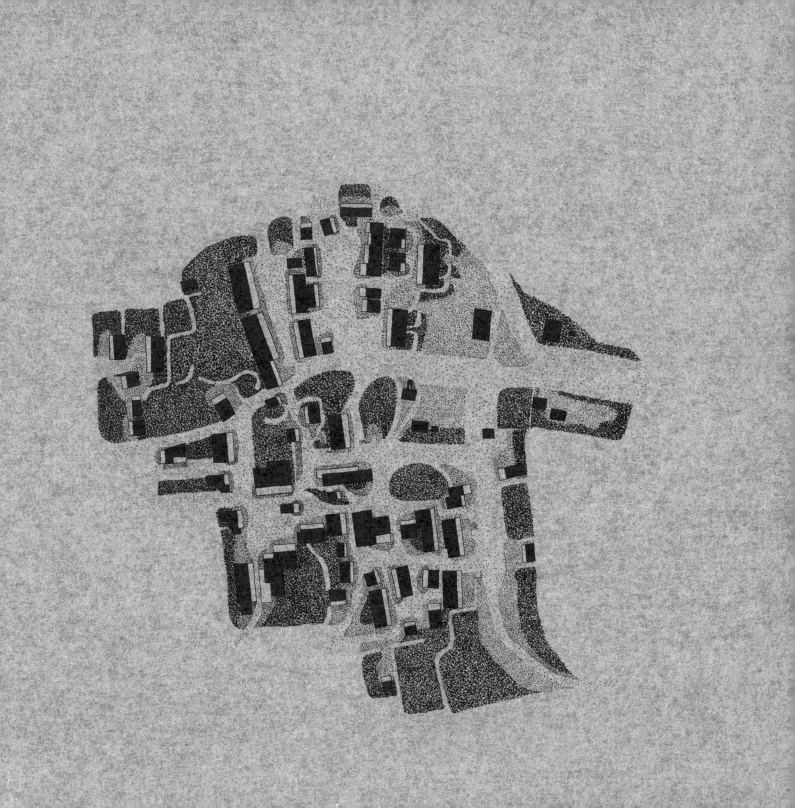

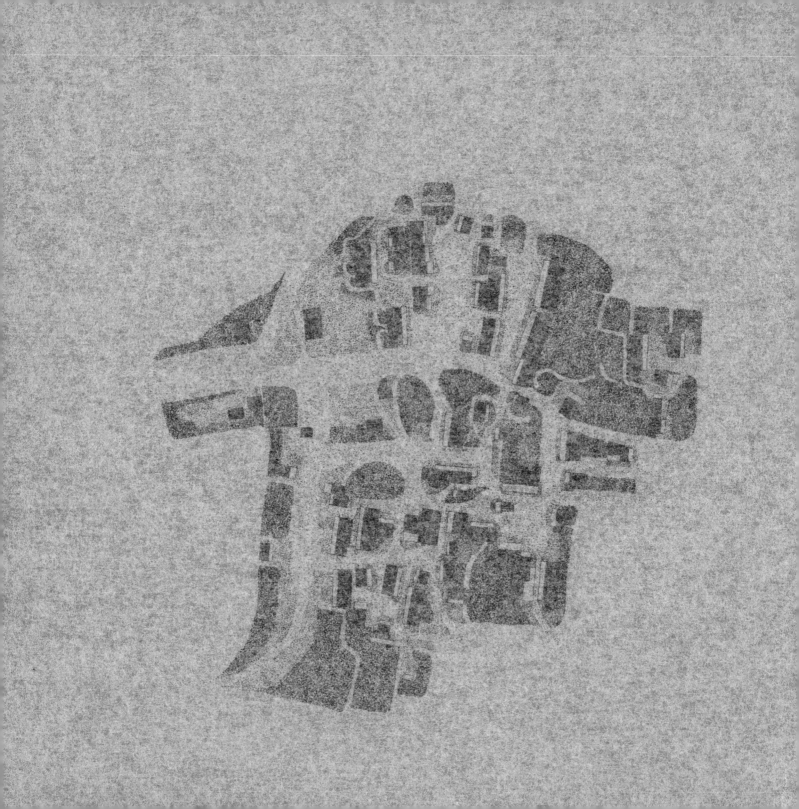

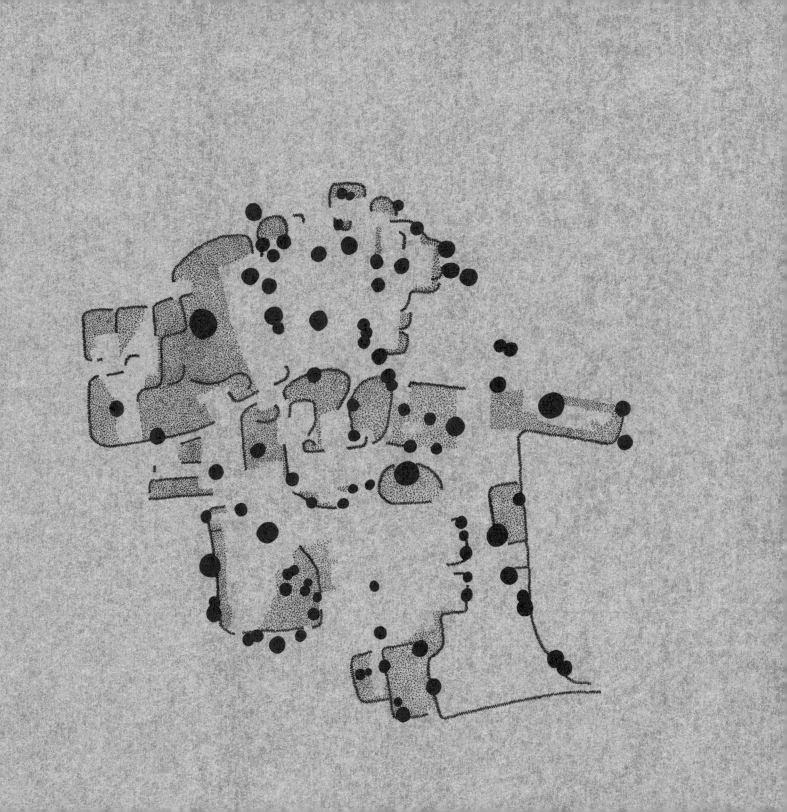

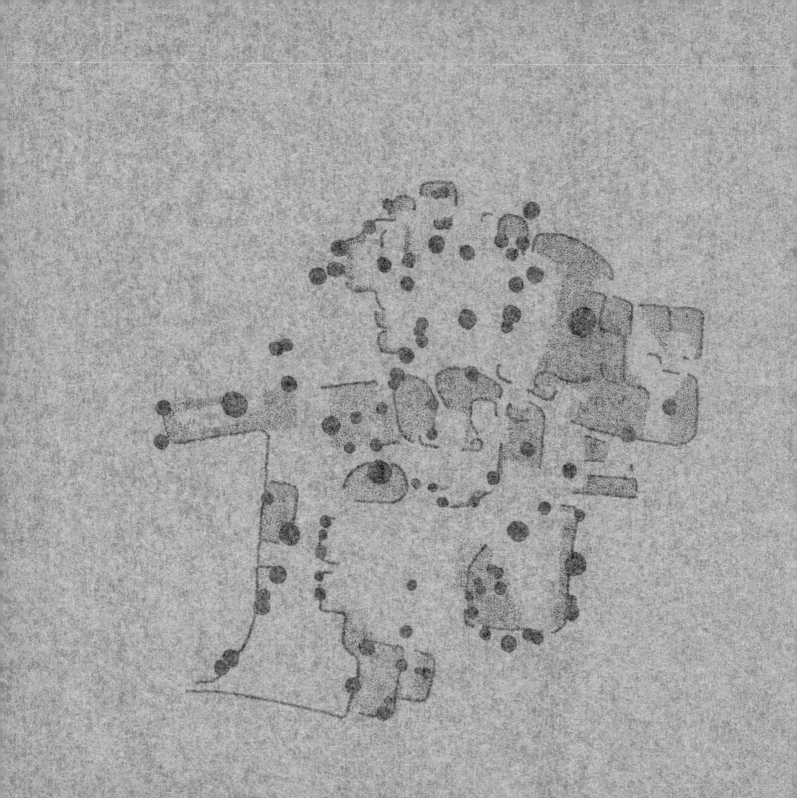

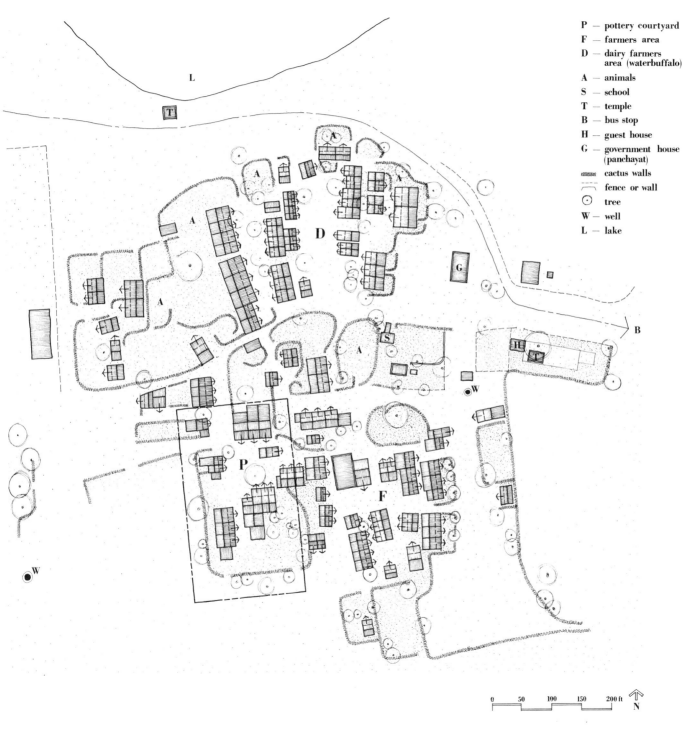

P — pottery courtyard
F — farmers area
D — dairy farmers area (waterbuffalo)
A — animals
S — school
T — temple
B — bus stop
H — guest house
G — government house (panchayat)
▨ cactus walls
— — — fence or wall
⊙ tree
W — well
L — lake

0   50   100   150   200 ft

N

prising roughly 50 percent of the village), agrarian farmers (35 percent), and pottery makers (15 percent). Family life and religious orientation is strong within all three castes, and many families pray at the temple six hours each week. Of greater importance, however, is the social seclusion of each caste. There is no communication beyond the formal ones of education and government (thus the *panchayat's* location "outside" the real form of the village).

Each of the castes is separated into its own area, and roughly organized around a courtyard—although in keeping with the general character of indigenous village growth, the courtyards are not sharply defined (and "activity area" would be a more accurate term). As one begins to understand the caste organization, an ordering element appears in the village. The dairy farmers occupy the northern half of the village. Their individual "yards" are larger so that their cows may be penned in during the night. Since grazing land is to the northeast and northwest, their location near these fields (and closest to the lake) makes sense. Although the houses of this caste are loosely grouped around a central activity area, the place is not of great importance, as the men are normally in the fields with their buffalo herds.

The farmers' caste occupies the southeastern segment of the village, closest to the grain fields. The potters' caste occupies the smallest area and is grouped around the best-defined activity area, or courtyard. Since the potters do not leave the area during the day (other than to go to the lake for clay), their courtyard is of greater social importance to their way of life. As mentioned earlier, outdoor and communal life dominates over the private. Overlay 2 to the general plan (illus. 3) further illustrates this point. Showing the public-to-private hierarchy of spaces, this overlay is in effect also a circulation plan, since all villagers and their animals communicate and travel freely through the public square. The arrangement of the houses into semiclusters around or near these activity areas also facilitates visits and talk between neighbors.

Since the house interior is rarely more than an enclosed storage space (where women may sleep during January and February), the verandah is used for cooking, washing, gossiping, entertaining, and resting.[5] This "outdoor" roofed room is the principal living space of every family in Vastrapur. The family cow is always kept close-by, and it is not unusual to see cow dung patties, formed by hand, drying on the side of a tree. These will later be used for fuel —every function of the water buffalo cow is utilized.

The hub of social life and the prime architectural element of the village house in India, the verandah's critical role can be seen by investigating in depth one caste, the pottery makers, and how they live around their courtyard.

5. *Verandah*, like many other Indian words, has slipped into English (usually with the *h* dropped, to become *veranda*. Other Indian borrowings include: bungalow, jungle, mufti, dungarees, sandals, teak, chintz, calico, shawl, punch, thug, loot, pyjamas, shampoo.

The Potters' Courtyard

The potters are the most closely knit caste within Vastrapur village, not only because they are the smallest, with seventeen families, but also because the making of pottery keeps them organized for daily work in and around an open kiln within their activity area.

The principal source of income for the potters is the making of large water-carrying (or storage) pots. These pots are almost identical in size and shape (fifteen inches in diameter) despite the fact they are hand-formed by twenty different male potters. Average wage figures are difficult to obtain, but are unlikely to exceed 1200 rupees ($240.00) per year for each working male, or approximately 3.5 rupees ($.60) per day.

Only bare essentials are used by each family, yet great sums of money may still be spent on important occasions. A daughter's wedding may easily cost 1800 rupees. Since no potter's family possesses this much ready cash, the moneylender is always used, and exorbitant interest rates of 50 percent a year are not uncommon. Although a few of the men supplement their incomes by working part-time as peons in Ahmedabad, in general the potter's family remains exceedingly poor.

All of the potters' families are Hindu and pray daily in the temple nearest the lake. They also observe the Hindu calendar, which has no specific weekly day of rest, but rather three holidays each month, coordinated with the full moon, the night of no moon, and the new moon. During these periods families often travel thirty to forty miles to arrange marriages for their children (within neighboring potters' castes). Although marriages within the immediate court-yard caste are discouraged, the ten families that actually face into the courtyard are all blood relatives. Approximately once a month the potters journey into Ahmedabad to see the movies or shop at the big market for essential food-stuffs.

The verandah is the hub of family life, the courtyard of caste life. The potters' courtyard is organized to provide for all facets of daily life—the open kiln for firing pottery, two large trees for shade, a low seating wall for resting and talking, and a storage area for drying and painting the waterpots. There is also a work area to the south where the twenty potters work together with their pottery wheels during special festival sessions, when small pottery jars and cups are used by all Indians, and the potters are particularly busy supplement-ing their normal income. During the morning, the courtyard is the scene of

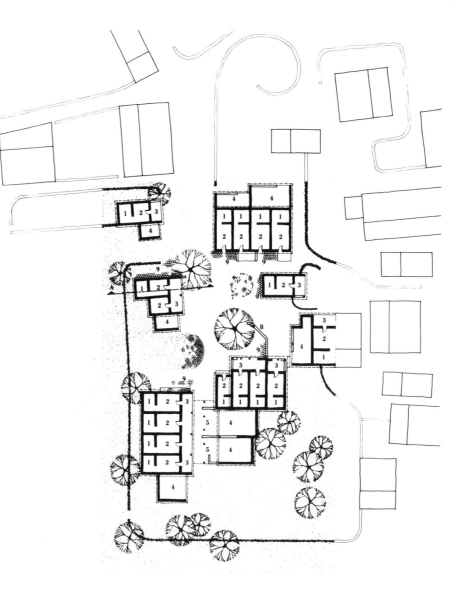

1 — storage room
2 — living space
3 — verandah
4 — workplace (no doors)
5 — special outdoor workplace with thatched roof
6 — storage mound (3'-0" high) for pottery
7 — open kiln area
8 — common seating wall
9 — pottery storage

4. *The Potters' Courtyard.* The most highly organized area of Vastrapur. Pottery is fired over an open kiln inside the courtyard space. Note special outdoor work space, covered with thatch, which is used by all the potters on special occasions.

teeth-brushing and family discussion; during the day, it is used by playing children, by cooking wives, by sleeping babies, by working men; during the night all the men sleep in the courtyard, while the women use their verandahs.

Two aspects of the potters' lives are worth special attention. First, each of the seventeen families owns at least one cow (five families own two). These three or four animals within it. The second feature is the attachment of a workroom to house groupings. These workrooms are used in beating out the clay and in storing half-finished pottery. The workrooms help explain the fact that four of the potters' families do not have verandahs (the only four such houses in the village). In these families, the workrooms are also used by the women, in lieu of verandahs.

To understand the relationship of courtyard and verandah to daily life, it is useful to outline the procedure used in making the waterpots that constitute the caste's principal source of income. First, mud-clay is gathered from the lake by all members of the family. This clay is stored in the workrooms and then mixed with finely chopped straw (usually by the women). The black clay must be kneaded and worked until it is just the right consistency. The potter then takes a lump of clay to the special outdoor work area (covered with loose thatch) where the potters' wheels are located. These large, very heavy stone and wood wheels are delicately balanced on a wooden peg so that they will revolve at great speed. After spinning the wheel by hand, the potter places his wet clay in the center, and as the wheel and clay revolve, he wets his fingers, squats beside the wheel, and proceeds to mold the water pot by hand. This may take fifteen to twenty minutes.

The wet, half-finished water pot is then carried back to the individual work area to harden partially. On the second day the water pot is worked by hand at the work area. Placing a stone inside the pot, the potter beats on the outside with a wooden mallet. Slowly the walls of the pot become thinner, and the size of the pot grows, while the clay hardens. After a few more days of curing, the pot is ready to be fired.

There is no furnace kiln at Vastrapur.[6] Instead, an open kiln is made in the courtyard about twice a month. First a few sticks are laid down, and these are covered with dried cow dung cakes until a fifteen-foot circle is formed. The unfired water pots are then placed all around the circle and on top, and "baked" or fired for about twelve hours. The open kiln method breaks about 30 to 40 percent of the pots, from too much, too little, or uneven heat. After the pots

6. The villagers recognize this as a problem (see Appendix) and would like a "gas-fired" kiln, although most of them are not sure what this means, other than that "it is a better way." Since a potter can produce only about four pots per day and gets twenty cents per pot when he sells them, a 25 percent breakage figure alone would mean that instead of eighty cents a day, he would make sixty.

are fired, the women and children paint them, either in the workrooms or on the verandahs.

While their husbands are moving back and forth from work area to court-yard to work area, the women of the caste are directing family life from the verandah. Here they cook over a small stove (the *chulla*, similar to a Japanese *hibachi*) using cow dung patties for fuel. They also gossip with neighbors, sing, grind wheat, wash the children, and work with the clay. Twice a day they go to the well for water. They also see that the children milk the cows and take them to the lake for water. The verandah usually faces the activity area, and it is always covered. It may have side walls, or more normally, half-walls (three feet high) which are used for sitting and for storing cooking utensils. At night, the women move their cooking utensils inside the house and drag outside the rough sleeping cots (*charpais*) used by men and women alike.

It is the interrelation of activity between verandah and courtyard that makes the potters' area interesting. As activities change, the area itself changes—from nights when the kiln burns and smokes and men work late, to quiet afternoons when babies are rocked to sleep in hammocks.

The physical form of the potters' caste area allows for little privacy; on the other hand, maximum social contact is needed to relieve the tedium of long days of repetitious work. In this sense the physical form relates well to the social need. Five hundred years of slow, minute change have evolved a simple, workable, indigenous, verandah-courtyard village.

Bapu's Family

Bapu, twenty-nine years old, and his wife, Sara, twenty-eight, live with their four children in a two-room house that forms the western edge of the potters' courtyard. The two oldest girls, Asha, ten, and Gita, seven, attend school, while the youngest girl, Indira, three, and the baby boy, Adi, one, stay at home, always close to their mother.

Their home is the major part of a larger structure also containing the one-room home of Bapu's brother, and a workroom where the two brothers work their clay pottery. Bapu's house area itself is absolutely typical of Gujerati village construction (see illus. 6). The back room is windowless and used only for storage of grain (for the cows) and for jars of peanut cooking oil, spices, and meal (for the family). The front room also is windowless, but since it has

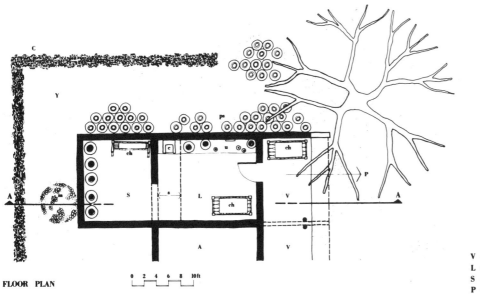

FLOOR PLAN

V — verandah
L — living space
S — storeroom
P — potter's common
     courtyard
Y — private yard
     space
A — adjoining house
C — cactus

c — chulla (cooking)
ch — charpais (cots)
s — shelf for storage
u — cooking utensils
ps — pottery storage
m — dried cow dung
     mound

5. *Plan of Bapu's House.* A typical two-room house, in which the verandah plays a critical role. Note the absence of windows. Light filters through the thatch roof during the dry seasons.

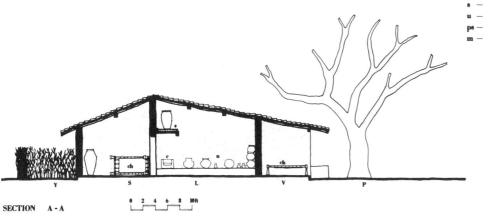

SECTION A - A

6. *Section of Bapu's House.* Note the high shelf for storage of cooking oil, etc., and the low shelf used for actual cooking.

7. *The Water Buffalo.* The villagers' means of survival. Almost every family owns at least one cow, using the milk to feed the children.

a wide—and in terms of village construction, handsome—carved wooden door, considerable light enters. In addition, the thatch roof is partially thinned during the dry months and light filters in from above. The entire home area contains less than 400 square feet (or 67 square feet per person).[7]

The "front" room, or living space, contains all the major implements needed for daily life—*charpais*, the light beds, which may be stacked; a few brass pots and pans for cooking and storing water; the small pottery *chulla* stove; a low shelf along one wall where the cooking utensils are stored; and a higher shelf which separates the front room from the storage space. On this shelf are the larger utensils, less frequently used. Beyond the living space is the outside verandah where Sara usually cooks, and where the family actually lives. Bapu's home is better constructed than those of many of his neighbors. The walls are built of kiln-dried brick; the thatch roof is thick and of good quality; the floor is new, of cow dung mixed with clay and water and allowed to dry to a highly polished sheen.[8]

By no means is Bapu wealthy. His total material possessions are his home,

7. This compares to about 500 square feet per person in an American subdivision, and 20 per person in the new housing developments of Hong Kong.

8. The cow dung floor is almost universal in the village. Once dried, the floor is hard, smooth, and lustrous. Until it dries, however (about three days' time) the stench is unbearable, and unhappy is the lot of a family caught with its thatch roof "open" during a sudden rain.

his bicycle, two cows, a potter's wheel, eight brass jars and pans, six cooking utensils, eight bracelets, and six anklets and three necklaces distributed among his wife and three girls. The total value of all these possessions is less than eight hundred dollars.

Everything Bapu owns is either in his home, in the immediate storage space around the house, or literally on his wife and children. The family's standard of living is materially better than that of many others. This is due to the fact that for two years Bapu has been working one half of every day in Ahmedabad as a porter (or janitor) at the university. This has enabled him recently to rebuild his home with help from his brother, and at this moment he is free of debt. The fact that he has three daughters, all of whom must be provided with dowries and expensive weddings, means that Bapu must watch his finances carefully. In the last four years the family has had only one piece of fruit to eat, and that was an orange which Bapu purchased for fifteen cents when Gita was sick. Still, he is not unhappy over finances—his income from pottery-making averages forty cents per day and his porter's pay fifty cents per day—and this is better than average. Indeed, so attractive is city work that six other potters also supplement their income in this manner.

The best way to understand the close interaction of social and private life with the courtyard-verandah architectural form is to study one day in the life of Bapu's family. The day begins at 5:00 A.M. when Bapu and his wife arise.[9] Bapu has slept on a *charpai* in the outer courtyard but near the verandah where Sara and all three daughters are also on *charpais*. The baby sleeps just inside the door of the living space. Bapu's first act is to brush his teeth with a stick, squatting in the courtyard, talking with the other men of the potters' caste. He then goes to the temple to pray for fifteen to thirty minutes. By the time he returns, Sara, who has brushed her teeth behind the house, has awakened the three girls. The boy, Adi, like most babies, is allowed generally to set his own schedule. Using the *chulla* stove, Sara heats some tea for the family. She says: "I've heard that sweetened tea tastes good, but I don't think I would like it; besides, I've never tried it. Generally our meals are simple. Bapu doesn't even like as many spices as most men." Sara has heated the tea (exactly one cup per member) using three twigs and half of a dried cow dung pattie for fuel. No food is taken with the tea. By 6:15 morning tea is over and Bapu heads for the "men's field" to take his toilet and a light sponge bath. Sara and her daughters, Asha, Gita, and Indira have gone to a separate field for their toilets and baths.

By 7:00 Bapu has joined his brother in the work room of their double house.

8. *A Small Child.* Wearing jewelry, this girl is typical of Vastrapur's children. The jewelry will form part of her dowry.

9. This data was gathered during a pleasant day (75 degrees) in late October of 1964. I am indebted to Miss Gita Shah, then a student in the Architecture School, who served as a translator for all conversations.

This morning they are "beating out" pottery and not using their potters' wheels. "When we use our wheels we usually work with the other men in the common workspace off the courtyard. This only happens about once a week as we only have enough fuel to use the kiln that much." The potters all use the kiln independently. The laying up of the sticks, cow dung, and water pots requires considerable skill, and while breakage may be as high as 70 percent, Bapu's average is about 25 percent. The potters help each other when in need, particularly when children are involved, nonetheless, certain skills, such as build-

9. *A Pottery Market.*

ing a good kiln fire, are still sources of pride, and no potter would think of sharing a kiln fire. Bapu and his brother, working as a family team, usually fire about twenty pots at a time.

From 7:00 to 10:00 Bapu works steadily on his water pots. During this time Sara and the girls milk and then feed the two cows, the girls taking the cows to the lake for water. On their way home the girls gather more soft clay, which Sara then stores at the rear of the workroom, watering it down and mixing some straw with it. Sara has also fed the baby, Adi, and gone to the lake herself to wash clothes with all the women, talking most of the way.

From 9:00 to 10:00 the girls play in the courtyard while Sara cooks the breakfast of fried bread and a few vegetables with spices. Two or three times a week, small pieces of mutton are added to the stewlike mixture of vegetables, but the family never eats fruit. The children get some milk with their break-

11. *A Typical Home in Vastrapur.* The verandah of a house in the farmers' area of the village. Note the *charpai* (cot) standing on edge. Almost everything the family owns is shown in this photograph.

10. *Main Street in Vastrapur.* This sandy walkway, defined by building edges and cactus walls is the main pathway for the village.

12. *Waterpots.* The principal source of income for Vastrapur potters.

13. *Bapu at the Pottery Wheel.*

fast, which begins at 10:00. By 10:30 Asha and Gita are off to school (which meets from 11:00 A.M. to 2:00 P.M. and again from 3:00 P.M. to 5:00 P.M.). Bapu eats with his family on the verandah and then rests for about twenty minutes before bicycling to the university where he works from 11:00 A.M. to 7:00 P.M. He usually carries with him a few wheat pancakes which he eats at about 2:30 or 3:00.

At 11:00 Sara washes her pottery dishes, using sand and a little water. From 11:30 to 12:00 noon she works with clay, wetting it, and checks the water pots which Bapu has stored, making certain they are not drying out too rapidly. She also paints two water pots which have already been fired and are almost ready for sale. Since 9:00 she has not walked more than thirty feet from her verandah.

She has cooked a meal, watered clay, worked with pots, fed her baby, washed dishes, prepared a light lunch for Bapu to take, and carried on three conversations with her sister-in-law, besides making comments to passing women. She says, "Sometimes on special holidays I cook some sweet things for the children, but most of the time, today is similar to other days. Of course, I like to talk with all the women. Today I learned that Krishna [one of her neighbor's children] has just been betrothed with one of my cousins in a close-by village. It will be a wonderful wedding."

From 12:15 to 1:00 P.M. Sara rests on her verandah, and then she goes twice to collect wood with two of her friends. Usually she also goes to the well, but today she doesn't need water. The youngest girl Indira, has been napping from 12:00 to 1:30, inside the house with the baby. Between 1:30 and 2:00, Sara cooks another meal of vegetables and *chappatis* (wheat meal pancakes) for all three daughters. Promptly at 2:00 Asha and Gita come home for lunch. This meal is eaten, as all meals are, squatting, the group roughly in a circle on the verandah.

At 3:00 the older girls go back to school. Sara then feeds the baby again, plays with Indira, visits in the courtyard with other women, and kneads clay until 5:00 when all the children gather and talk and Sara joins in. The children finally take the cows to the lake again and Sara begins the evening meal. "Everything I need is right at hand," she explains. "I'm so pleased with this new house which Bapu and his brother built, particularly the ledge in the living space which holds the *chulla* and my other cooking things. Our old house didn't have this and it was difficult to keep things clean."

At 7:00 Bapu comes home from the university. He works on his pottery

14. *The Potters' Courtyard.* The edge of Bapu's house is on the extreme right, the open kiln at the extreme left. Bapu's two cows and one his brother owns are tethered in front of his verandah.

15. *Pottery and Houses.* Along one edge of the courtyard, the waterpots are lined against the houses, ready for painting by the women.

until 8:00, when the family has its last meal together, and they all talk about what happened during the day. This is a leisurely meal, and the children are put to bed as soon as it is over.

From 9:00 to 10:00, families set up their *charpais* on verandahs and in the courtyard. Bapu and his wife walk across the courtyard to discuss Krishna's betrothal with their friends. The courtyard is filled with a low murmur of conversation. The kiln is burning, and one potter is using it. Since it is a work-day evening, no music is heard, only a lengthening of quiet, until at 10:00 the village is silent, and everyone in Vastrapur is asleep. The day ends much as it normally has for some five hundred years.

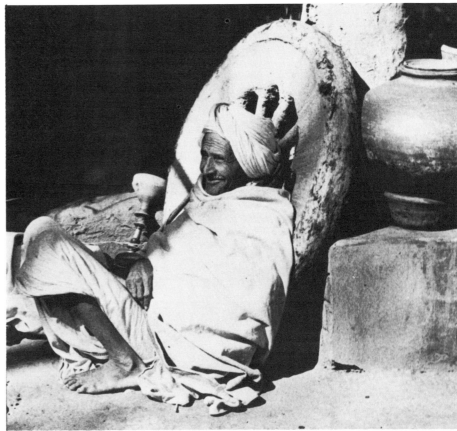

16. *Old Age in Vastrapur.* A retired potter, respected by all the younger potters, basks in the sun in the potters' courtyard.

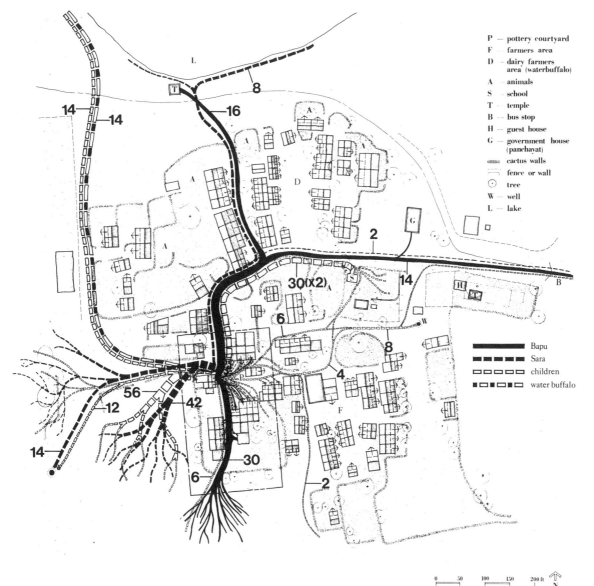

P — pottery courtyard
F — farmers area
D — dairy farmers area (waterbuffalo)
A — animals
S — school
T — temple
B — bus stop
H — guest house
G — government house (panchayat)
⊞ cactus walls
═ fence or wall
⊙ tree
W — well
L — lake

Bapu
Sara
children
water buffalo

0 50 100 150 200 ft
N

17. *Activity Map.* This map traces the pathways taken by Bapu's family and cows for one week. The frequent trips just outside the village reflect the use of the toilet fields, one for women and one for men. Note that the family never travels through the farmers' courtyards. While the path pattern is casual and the family may travel wherever it desires, there is in fact no need to go through the center of another caste's activities.

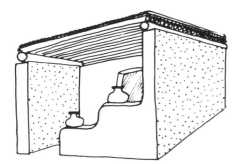

18. *Verandah.* The typical Indian veran-dah, a half open space. This particular ex-ample uses a stepped wall to create a useful shelf area (it may also serve for seating).

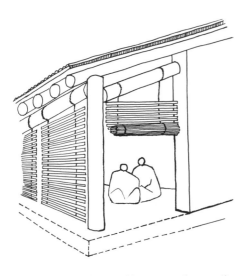

19. *Bamboo Mats.* Mats are frequently hung around the verandah edge to serve as screens. This is a simple but effective way to control the sun.

This day has been a typical one for Bapu's family, but since he works in Ahmedabad their weekly schedule is slightly different from that of many other villagers. One day a week (on Sundays) he does not work at the university. Since his children's school is also on a six-day schedule, Bapu likes to take his entire family with him to a market at the edge of Ahmedabad on Sundays. On the holidays, when everyone in the village goes visiting or into Ahmedabad, Bapu's family does the same, although Bapu does not enjoy the movies be-cause of minor eye trouble, and therefore doesn't go. He enjoys shopping and visiting the temples in the city. He is deeply religious, but this is not uncommon with village men. "I find the [Hindu] temple a restful place and I give thanks every day that my children are all well, that I have a good job, and that my wife is good to me."

He is looking forward to the electrification of Vastrapur—to television and to the fact that his children will be able to read to him at night. (Lines were being brought to Ahmedabad as this study was being conducted.) Three photos (posters) hang in his living space: "They are of the Mahatma [Gandhi], Premier Nehru, and President Kennedy. These men were for peace. With peace we can learn to solve India's problems. We can become great again, as a country. But more than this Vastrapur can be a better place to live."

Conclusions: The Quality of Village Life, and Implications for American Housing

Two general conclusions emerge from this anthrophysical study of Vastrapur. The first of these is that caste social life predominates over family private life. The corollary is that the physical form reflects this social emphasis. That is, given several hundred years of growth, the caste system (or social force) has molded the physical form (the architecture) to fit the social-functional need. (Given time, "form follows function" in the broadest signification of that famous statement.) This fact was demonstrated by the general circulation system, the public-to-private hierarchies of space, and by the lack of privacy-oriented forms. The diagram of Bapu's family activities for one week further confirms this conclusion. Most significant, however, is the way Bapu and his family use their verandah and the potters' courtyard for all their daily activi-ties. The verandah is the focus of family life, the courtyard of caste life—and both are open and community-oriented.

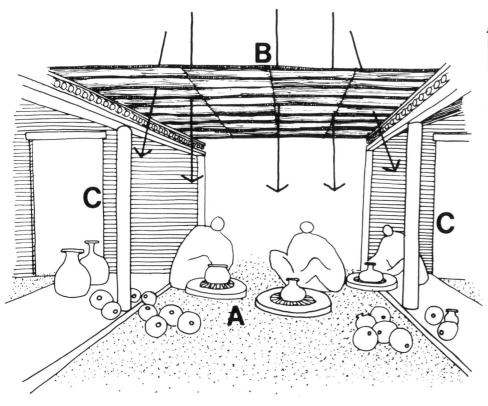

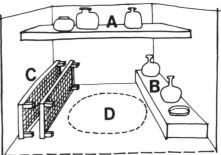

20. *Principal Room.* A typical layout for the major room of a villager's house. (A) The high shelf used for storing oil, grain, etc.; (B) the low shelf used for cooking; (C) beds (*charpais*) stacked on edge to save space; (D) the woman's work space.

21. *The Workplace of the Potters.* At the southwest corner of the potters' courtyard, a workplace extends down between potters' verandahs. This place (A), where all of the potters frequently meet and work, is protected by a loose thatch and bamboo roof (B). The verandahs (C) are used for storage of clay, water, and newly made pots.

The second general conclusion is that the shaping of transitional places is vital to the success of Vastrapur's form. The variety and subtlety of these transitional forms cannot be overemphasized. For example: cacti walls which define private places (in function) but allow visual communication; thatch roofs which when thinned allow filtered sunlight into the home interior; verandahs that are covered but open (semi public); low walls on which villagers may sit, but which also define semiprivate space; the potters' work area, adjoining yet separate from the courtyard and covered with loose branches to filter sunlight; the relation between the few large trees in the village and the clustered houses, the lower cacti walls, the moving humans and animals, and the open fields leading out to the toilet areas—the list of such transitions (both social and physical) is almost endless. They testify to a long civilization, many-layered, and far more complex than a casual visitor would readily perceive.

As to the separate question of what changes must occur when Vastrapur is engulfed by Ahmedabad, the answers lie outside this study (although an attempt is made in the Appendix to wrestle with the problem). Certainly electricity will affect village life. Close behind will be the car, plumbing, sanitation, and a different work schedule.

What can we learn from Vastrapur which may be applied to United States housing problems? It is dangerous to look for specific lessons. Two points however, appear to merit further consideration in relation to ground-level housing in the United States. The first is that the meanings of modern life should be appropriately reflected in the physical form of our neighborhoods, rather than having physical form attempt to impose a style of life on us. For example, there is nothing inherently wrong with barbecue cooking, if suburban (or urban) dwellers find it pleasant. But to institutionalize this pleasure in development houses (by building a permanent barbecue) deprives the future owner of flexibility. Again, even if a house in the form of a box is cheap to build, the shutters might at least be left off, and the purchaser allowed the choice of applique (or better yet, the choice of taking the money value of the shutters, and putting it into a two-foot longer box). This point will be explored more fully in the last chapter. In Vastrapur there is no conspicuous waste.

The second point of application from Vastrapur's form is its concern with transitional places. An understanding of the importance of such ordering elements might make our new housing far more livable. Vastrapur relates to the walking human, the water buffalo, and the meandering and playing children, by means of a vast series of subtle transitional elements. Why cannot we begin to acknowledge some of our obvious transitional problems, such as where to park the car,[10] how to get out of the car and walk (covered) into the house with groceries, how to find the real front door, how to take off a raincoat and walk to a bedroom without going through the middle of the living room (or family room, den, parlor, play space, study), how to allow children to play outdoors without being run over, how to sunbathe in privacy, how to sleep in privacy, and so on. All of these questions involve movement through transitional places, screens, walls. This point, like the first one, will be explored

10. A number of American upper-middle class suburban cities have solved the "car problem" by zoning it off the street; that is, no garages may face the street, but must be entered from the side or the rear. The results of such restrictions are predictable—houses often contain a circular front drive as a guest entry, and a car (usually the homeowner's) is parked there, while the "service drive" invades the rear yard areas.

further in the last chapter. It is sufficient to state here that the principal lessons from Vastrapur are that its physical form works well because it relates to the human needs of its inhabitants.

Appendix: A Problem Given to Architecture Students in Ahmedabad, and Two Solutions

This appendix consists of the notes, program, research data, and solutions to a problem given to third-year architecture students in India, based on the assumption that the resettlement of Vastrapur on a nearby site would be possible with government assistance. The program was suggested by several considerations: the village would be engulfed shortly by Ahmedabad, such a resettlement effort lay within the sphere of then current government interest and policy; and the study of Vastrapur would in itself increase the familiarity of beginning architecture students with basic social patterns (none of the students was from a village environment).[11]

The problem was approached in three stages: 1) investigating village form, including the caste system and its influence; 2) defining those changes that would be necessitated by resettlement, then those changes that would be desirable, and out of these formulating a "new village" program; 3) designing a village plan to conform to this program. Conclusions from the first stage have been generally included in the preceding case report.

Assuming resettlement to be possible, during the second stage of study the students isolated certain changes as mandatory from a governmental point of view. Only a summary is included below, not the endless discussions preceding the goals. For example, students decided that urbanization had already made an impact on village life—a weakening of religion, an increase in educational level, an increase in material possessions, and an expectation of improvement. Nonetheless, students also argued that if it *were* possible violently to change village values for an urban outlook, then the quality of present village life would deteriorate without a gain in understanding of possible advantages in urban life. Beyond this, students also argued that such a violent change would *not* be possible—villagers living in a semirural area would not give up their cows. For this reason, all students agreed that imposed changes should be

11. It would be unfair not to add that studying Vastrapur also provided a new learning experience for me. The best teaching occurs when the faculty member is also learning—albeit he brings to the problem a greater expertise. Pedanticism begins when the joy of discovery ends.

minimal and preceded by intensive social training, with maximum flexibility in design for future improvements. The changes deemed necessary were the following:

1) Standardized, more permanent construction would be used—probably brick and concrete but consistent with minimum economic investment.
2) Water and electric power would be supplied at every house.
3) Some form of central sanitation system would be employed.
4) Only one central road would be paved.
5) Regular bus service would be maintained.
6) The *panchayat* and school would be relocated, with free space for one temple.
7) The village would be expandable.

These further changes (or goals) were considered desirable.

1) The relation of village life to the lake should be reemphasized.
2) The unique relation between families and their cows should be preserved.
3) A central space for all the village should be more clearly formulated (and include television reception).
4) Houses should be so situated as to maximize breezes, to control sun, to facilitate group communication.
5) Houses should have at least one private area.
6) The verandah concept should be utilized.
7) A new kiln should be supplied for the pottery-makers.
8) An intensive training period of from one to three months would be mandatory before moving—that is, villagers would have to be trained in the use of toilets before any sanitation system would work.

Out of this study emerged a condensed program based upon the following general statement as to design approach:

It is the obvious attempt of the government to disrupt village life as little as possible, while helping to improve the mental, physical, and social life of its people, and to prepare them for emerging urban (and national) values.

The problems of village design are diverse—even within the sphere of the architect or planner designing the "framework" only. In India the general poverty and meager existence of the lower classes must be changed—no other social problem is more important to the well-being of the country as a whole. Architectural resolutions of this issue in terms of housing must recognize, however, that "quality" is dependent on more than plumbing. The concept of living which each designer embodies in his plan will be the critical issue. The attitude followed should be to develop a working design. Students are urged to forget for the moment "fashion," sculptural architecture, and examples easily copied without understanding the program behind them. To recall the statement one of them made during their research: "To see the cow at the front door is considered a good omen."

Several requirements for design of the new village were taken to be mandatory:

1) All statements as to goals and changes discovered in prior research are to apply to the design.

2) On the site shown (illus. 22) design a new village to consist of 100 houses: 60 units designed for four occupants and 40 units designed for six occupants, with expansion possibilities.

3) Employ a "rowhouse vault" construction system, making maximum use of brick and minimum use of concrete (illus. 23). This system is common around Ahmedabad. Concrete floors will be used. Each house is to have water, electric power, and plumbing.

4) Provide for some form of minimal village center.

5) Consider maintaining the caste system but also consider its eventual disappearance.

6) Plant fifty trees donated by a mill owner, locating each tree carefully.

With these guidelines, the students worked to develop a village plan. Only one of their solutions is included here (illus. 24, 25). Developed by Miss Gita Shah, this design contains several features worth consideration.[12] First is the basic house plan itself. Using a two-vault-wide unit, this plan leaves one vault completely open, with a free-standing storage closet in the open through-space. The storage closet acts as a space divider. The front part of the covered area

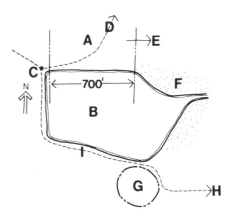

22. *Site Plan for New Village.* This site plan was used in planning a new village layout for Vastrapur. (A) New area for village, facing lake and seven hundred feet wide; (B) the lake; (C) bus stop; (D) path to neighboring village; (E) farming fields; (F) flood lands, to be used for grazing; (G) the present village; (H) to Ahmedabad; (I) roadway for bus.

12. Miss Shah's actual drawings have been redone to fit the format of this book.

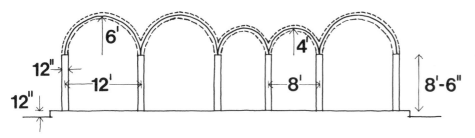

23. *Village Structural System.* This system was used by students in planning the new village. Walls were assumed to be twelve-inch bearing walls made of soft brick, with vaulted brick roofs, either eight feet or twelve feet in diameter. This is a common building system in Ahmedabad.

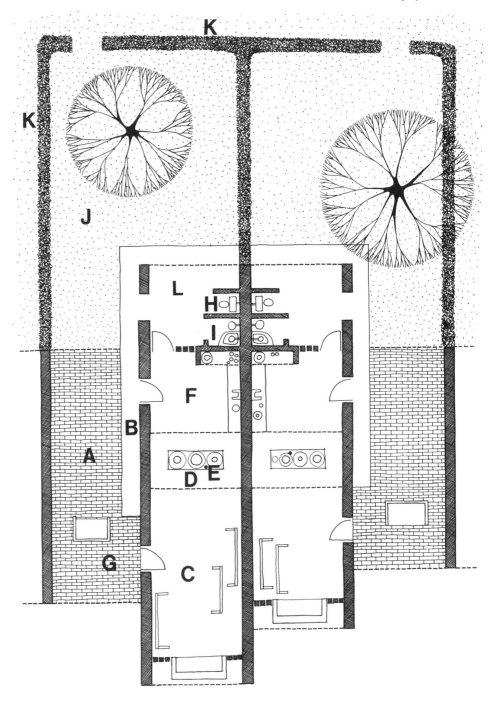

24. *New House Plan.* This student solution ingeniously preserves many amenities of present village life while modernizing the unit plan. The verandah is replaced by an open-ended vault, broken into two spaces (A and G) by a movable storage unit. The woman may use either room (F or C) for cooking or storage. Air circulation is assured by screens, front and back. The high storage shelf (D) has a light, movable, wooden storage unit below it (E). Toilet and water (H and I) are outside the house in a separate work area (L). The Cows may be kept in the back yard (J) or in one segment of the verandah. A cactus wall (K) surrounds each unit.

25. *New Village Site Plan.* This solution, utilizing the house plans shown before, also preserves village cultural values. Houses are loosely grouped around courtyards, with the school and government center (*panchayat*) in the center of the village. Note that expansion of such a village would be simple.

KEY
A   Bus stop
B   Road to neighboring village
C   Farmers' entries to fields
D   Houses
E   Entry to school
F   Back yard cow pathways
G   Loosely defined courtyards
H   Government center (*panchayat*)
I   Commons, or green space
J   School
K   Playground
L   Possible future store locations
M   Lake
N   Paved road

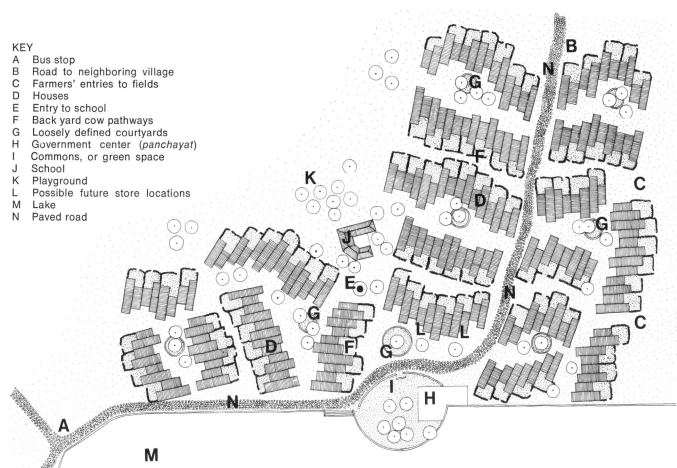

may serve as the verandah and the back as a "cow space" (or, in the case of potters, a work space). The open vault also allows the cows to pass from front to rear of the house, so that each animal may use a water outlet attached to the rear of the house. Further, Miss Shah's scheme recognizes the difficulty of training villagers for toilet sanitation by placing this fixture outside the house unit.

Inside the main living space vault, she has rearranged the typical village house only minimally, keeping a two-room division, separated by a loft space, with traditional storage areas. The majority of students and faculty agreed with this approach, emphasizing that the villagers' way of life should be disturbed as little as possible. The six-person houses are of course longer than those designed for four persons. In order to achieve variety and "courtyard" village organization, the houses are arranged in an informal manner along loosely defined streets. This arrangement makes for easy expansion, while maintaining the varied character of the older circulation pattern.

The importance of the lake is emphasized by locating the village center at the edge of the water and having the one paved road pass through at this point. The fifty trees are concentrated within the irregular courtyards, giving further emphasis to these meeting places. Considerable insight is brought to this solution, and it is a remarkable design for a beginning architecture student. It has a quality which, above all, is Indian in character.

An interesting question remains: How could such a *village* scheme be sold to a national government primarily concerned with burgeoning *urban* housing problems? Village architecture is possible only if it can be shown to improve living conditions (a social issue) and if it can be constructed with a minimal capital investment (an economic issue). Miss Shah's solution admirably satisfies the first condition, the second requires further thought. The outlay for water and sewer lines represents one of the major (and one of the first) capital expenditures. Past experience shows that developers (private or governmental) minimize this investment. Accordingly, the solution was restudied with new guidelines: to preserve the quality of the design but to simplify utility lines as far as possible (while still leaving water and sewer connections at each family unit). This called for straight-line utility connections of minimal distance and easy maintenance.

If water and sewer connections were removed from the houses entirely and placed at the rear lot corners (illustration 26) it would be possible to have a straight-line utility system.[13] If this system were further refined so that it lay

13. These revisions both to the basic unit plan and to the total village plan, are mine.

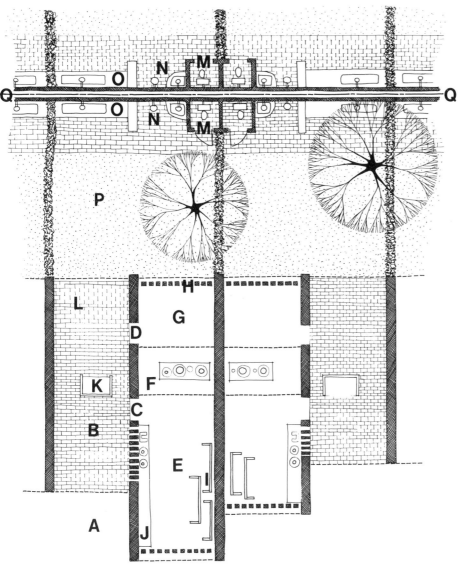

KEY

A   Entry from street
B   Front verandah (living)
C   Main door
D   Optional door
E   Main living space
F   Loft (above) with storage below
G   Storage room (or living, if desired)
H   Brick screens
I   *Charpais* (cots) stored on edge
J   Low shelf for *chulla* stove
K   Movable storage unit
L   Cow's space (or work place)
M   Toilets (four adjoin)
N   Sink (tap) and shower
O   Water trough for cows
P   Yard
Q   Utility line

26. *New House Plan* (*Revised*). The water and sewer lines are farther removed from the house, reducing initial government investment and making maintenance and supervision easier. This plan also allows for a watering trough for the cows. The house plan has been only slightly changed to improve air circulation.

within a double wall (or double cactus hedge), regular maintenance checks would be simple. Separation of the toilet from the house proper seems desirable for ventilation, while a watering trough for the cow then becomes possible. Applying this concept to the total village plan (illustration 27), one can see that the imposition of a utility framework does not have to mean loss of variety in spatial organization.

Both solutions emphasize two ideas: that Vastrapur's present village pattern corresponds well to the way of life of its people, and that the new pattern (or physical form) should not destroy what works. There is no need to make New Vastrapur self-consciously beautiful. The elements of beauty are already there.

KEY

A Bus stop. Paved road through village begins here, as does utility line.
B Utility line runs down center of road
C Lake
D Shops
E Planted trees
F Plaza
G Temple
H *Panchayat* (government house)
I School
J Houses
K Courtyard
L Back yards
M Utility lines (all water and sewer connections)
N Extension of open space
O Paved road to next village
P Possible change in direction of utility lines
Q Tree park
R Reserved for future use

27. *New Village Site Plan* (*Revised*). A site plan utilizing the concept of the student solution above, but reflecting minimal initial government capital expenditure. Expansion need not occur in a straight line, but might bend, as shown by (P) dotted.

IV Conclusions

The Anthrophysical Methodology: Advantages and Disadvantages

THE greatest advantage of the anthrophysical method is that it is an uncomplicated means by which architects and planners can investigate housing form. By concentrating on one family and its relationship to the land, the interviewer can isolate those determinants of form which are most effective for a specific neighborhood.

The investigations lead to suggestions for further study, rather than isolating truths; but perhaps more important, an assemblage of case reports yields interesting comparisons, not merely about neighborhood form, but also about different (and changing) social attitudes toward form.

The greatest strength of this method is also its inherent weakness. It assumes a prior awareness of how architectural form is shaped, and a knowledge of basic interview techniques. This does not seem an overwhelming obstacle, however, and given intelligence and desire, it should be possible to collect a series of case reports relatively easy to compare.[1]

The method has several other advantages not often found in architectural research. First, it is possible to put the conclusions in a written and graphic form comprehensible to laymen.[2] This is important when the present tendency is for each profession to move toward ever smaller and more technical areas. The approach also allows conflicting data to be brought together and related. For example, the pathways to work taken by Bapu in Vastrapur and by Peter Steinman in Parkview, are *not* the shortest pathways measurable on the map (Bapu found his route more enjoyable; Peter *thought* his route was shorter). In each case the anthrophysical approach encourages immediate questioning and confirmation of such choices.

Finally, the approach lends itself to team work and to data-gathering by more than one group, increasing its potential for assembling and then comparing a series of case reports. Moreover, the conclusions could easily be stored in computers, further facilitating comparison. In order to avoid the dangers of over-simplification, the method needs to be refined, so that all data gathered can be checked and reevaluated if any conclusion is in doubt. This work is now in progress.

---

1. Utilizing team teaching with a sociologist and a land planner, I have found third-year college students (in architecture) capable of competent data gathering and evaluation.

2. I am aware of the great dangers of over-simplification. Unfortunately these dangers are present in any approach. Hopefully we are no longer so naive as to believe there are simplistic panaceas to any problems of urban life.

Comparison of Case Reports

There is an interesting correlation between social attitudes and housing form in the two reports, both of which, significantly, reveal family satisfaction with the housing and the environment. The individual, fenced lots and private houses of Parkview correspond perfectly to the symbolic dream home of middle-class America.[3] The verandah-courtyard life of Vastrapur likewise correlates with the caste system of India's villages. In a study of Ichinomiya, Japan, not included in this book, such a correspondence was not found. The reasons for this were complex, but they involved building a new satellite neighborhood without provision for traditional amenities or for the rich mix of Japanese village life. As a result, the Japanese families were not happy with their environment.

A second observation is that the families interviewed stressed (explicitly and implicitly) the need for several ordered systems of form. I suggest initially three systems of importance: circulation, scale, and continuity/variety. All families interviewed (including the Japanese family) understood and appreciated differences in circulation—in walking, bicycle riding, and motorized transport. They also responded to changes in scale, and both Parkview Place and Vastrapur contain a variety of nodal points at which such changes occur (see illus. 1 and 2).

While a system of continuity/variety is difficult to analyze (partly because the term *variety* itself implies change), it can be stated that family use of the neighborhood space is as important as the physical form of the space. In Park-

3. Amos Rapoport in *House Form and Culture* (Englewood Cliffs, N.J.: Prentice-Hall, 1969), states that ". . . builders and developers never build houses, they build homes. The dream home is surrounded by trees and grass . . . and must be *owned*. . . . It is not a real need but a symbol." And again, "The popular house is based on the ideal that one's home is indeed one's castle, and on a belief in independence."

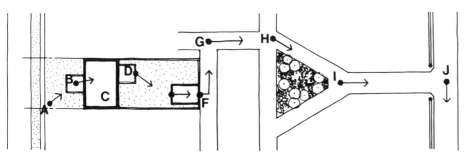

1. *Nodal Points of Parkview Journey.* A sequential interpretation of changes in scale: (A) Man leaves his front yard (semiprivate); (B) then goes from porch (semiprivate, roofed) into (C) house (private), then to (D) back porch (private, outside, and roofed), through yard to (E) garage, backs out car into alley (F); travels up alley to (G) a Parkview Place street, past park (H), (I); and through gates into major city arterial street (J). Each change in scale or turn in direction is clearly defined architecturally.

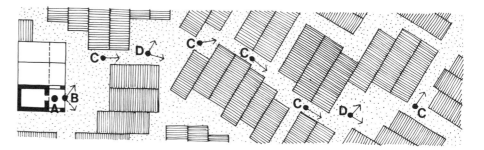

2. *Nodal Points of Vastrapur Journey*. A sequential interpretation of changes in scale: (A) Man leaves his verandah (semipublic), (B) then steps into courtyard (public); from this point on he walks through a series of public paths (C), and public courtyards (D). The differentiations of spatial change are subtle; they depend on moving people, dogs, cows, activities.

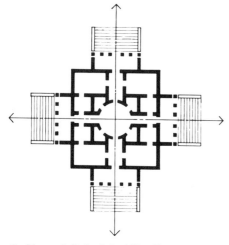

3. *Plan of Palladio's Villa Rotunda*. One man at the center of his villa dominates space in all directions. This conception, humanist in origin, has strongly influenced English and American architectural thought. Nowhere is it more evident than at Monticello, the home of Thomas Jefferson.

4. Susanne K. Langer's *Philosophy in a New Key* (New York: Mentor Books, 1958) contains a good analysis of symbolic transformation.

5. See Robert Venturi, *Complexity and Contradiction in Architecture* (New York: The Museum of Modern Art, 1966).

view, for example, the fact that the family normally leaves the neighborhood each day makes returning to it a change from the normal urban environment. In Vastrapur, the seemingly total conformity and continuity of housing pattern is constantly varied by an endless flow of people bringing news, activity, and color into the area.

Although there exists no precise terminology for these systems, people are nonetheless aware of the ordering elements in their environment and are constantly reacting to them. If the systems do not please them, the people, given sufficient time, will change the systems.

Carrying this point a step further, one can argue that an understanding of *how* systems of physical form are ordered is at the heart of the creative process in architecture. This creative process involves a synthesis of alternatives, never a simple elimination. For example, given possible design solutions of A, B, or C to a problem of physical form, the process of synthesis will never simply choose one solution over another (B over A and C); rather, it will be a creative statement which will contain A, B, and C—although it may rank B higher in importance, as in the formula X(B)(ac). This creative process must be symbolic because it deals with transforming into physical form the social and cultural aspirations of both the society and the artist.[4]

If it is true that society is always searching for an understanding of its ordered systems of form, then it follows that evolving theories of architecture cannot be built on complexity and contradiction, any more than they can be built on reduction.[5] Vastrapur and Parkview both clearly demonstrate the societal need for (and appreciation of) ordering systems in our physical environment.

Another observation of interest, noted earlier, is that the families think of their neighborhood as being organized by something other than the primary school. As Frederick Gibberd observes in *Town Design* (New York: Praeger, 1953)

The object of arranging the town's housing in the form of neighborhoods is to enable the family unit to combine, *if it so wishes* [italics mine], with other families into a community which has definite social contacts and a recognizable physical unity. The neighborhood is essentially a spontaneous social grouping, and it cannot be created by the planner. . . . A good shopping center . . . will like the medieval market place, bring the inhabitants into social intercourse far more effectively than any number of community centres. (P. 220)

A final comparison concerns each family's attitude toward the land. In the preface I suggested that an initial bias was the belief that families with children had to be induced to remain in the city, if the city as we know it is to survive. For this reason I chose to study environments containing house units with ground access, facilitating child play. It is obvious that no social or physical comparisons are possible with such slight data. It is equally obvious, however, that the question of social attitude toward land use is a central issue in planning new housing for any society, and that this issue is worth serious discussion.

Areas for Future Study

The most obvious area for future studies utilizing the anthrophysical method lies in continuing the case reports of existing environments, so that later comparisons will be of greater value. A further potential, and one of deeper significance, involves postulating physical variations in neighborhood structure, and then testing these variations with real families. Although this is a fascinating possibility, it would be no easy task. It is full of risk, because it involves a "trade-off game."

There are two general types of housing trade-offs, with no distinct division between them—those involving convenience, and those involving physical change. Trade-offs of convenience are concerned with manipulating cost com-

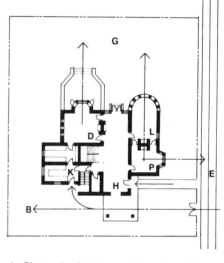

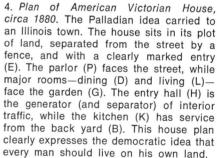

4. *Plan of American Victorian House, circa 1880.* The Palladian idea carried to an Illinois town. The house sits in its plot of land, separated from the street by a fence, and with a clearly marked entry (E). The parlor (P) faces the street, while major rooms—dining (D) and living (L)— face the garden (G). The entry hall (H) is the generator (and separator) of interior traffic, while the kitchen (K) has service from the back yard (B). This house plan clearly expresses the democratic idea that every man should live on his own land.

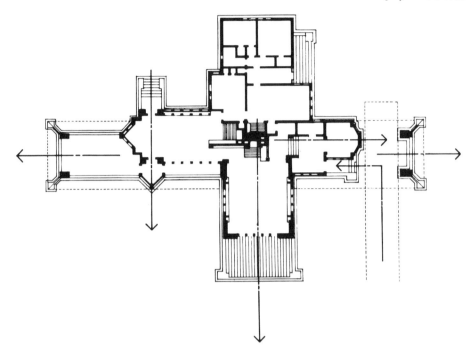

5. *Plan of Willetts' House, by Frank Lloyd Wright.* The logical extension of the Palladian concept in American terms, the "prairie house" fits to the land, extending arms into the landscape, making the land *itself* part of the house. Wright has replaced Palladio's central "man" with a fireplace. Now man faces inward to the hearth for comfort and security; he turns and moves outward into his controlled landscape. This concept is anticity, Transcendentalist, and peculiarly American.

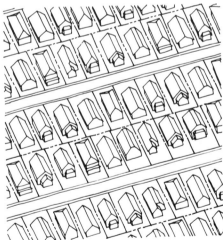

6. *Subdivision, circa 1947.* An illogical extension of the Palladian idea. The density is too high to support a separate house on a separate plot of land. The space between houses is wasted.

binations of air conditioning, humidifier units, shutters, panelling, kitchen appliances, the number of baths, bedrooms, basements, fireplaces, closets, fences, trees and so on. It would be quite feasible to use these elements to construct models that feature a variety of trade-off possibilities. This could be done first in model form, and then at full scale.

For example, housing form could be postulated on the basis of two extreme positions concerning convenience amenities, the first position being to maximize bedroom and living space in the house (trading off a great number of appliances, air conditioning, light fixtures, and so forth to make this additional space economically feasible). Position two might be an opposite extreme, for example maximizing land privacy and outdoor conveniences—trading off one bedroom and other indoor volumetric space to get fences, patios, outdoor lighting, swings, a pool.

Scale models could be built, showing a parcel of land developed in both manners, and the models could be evaluated through an anthrophysical approach. After ten to fifteen other "positions" had been postulated and tested, full-scale versions of the three or four most promising could then be built and retested in actual use. This enterprise would probably require considerable federal financing, or at least government tax relief to encourage private investment.

Trade-offs involving major physical change differ from convenience trade-offs principally in scale. In addition, they directly involve the developer's initial investment. For example, will the consumer accept joining houses with a party wall (i.e., rowhouses) if it can be shown that the utility lines will be buried so that more open space can be provided, and that the amount of paving will be reduced? Is it more important to identify the front and back of the house, or to concentrate all services (including guest entry) on one side of the house, leaving the back open to a park? Other elements which can be involved in trade-offs include tree planting, parks, off-street parking, common areas (pools, playgrounds), private yards, alleys, the size of streets, of sidewalks, of garages, and so forth.[6]

Trade-offs involving physical change are particularly murky business because they involve social and cultural attitudes toward land-use. We are currently clinging to a hoary Anglo-Saxon belief that land property must be "family-owned," and protected from its neighbor. Amos Rapoport has correctly surmised that this belief is a symbol and not a real need: "This symbol means a free-standing, single family house, *not* a row-house, and the ideal of home is aesthetic, not functional."[7] Whether Americans can retain this symbolic dream of territorial right in the face of population explosion, pollution, and disappearing land, is questionable.

In the illustrations accompanying this section I have attempted to show three different attitudes toward housing and land use: 1) land as private territory (as in England and the United States), 2) ground-level land as public domain with families living above it (as in much of Europe); and land used to group courtyard house clusters, often with public and private sectors mixed (as in Mediterranean countries and the Near East). There are, of course, other attitudes, but these constitute a firm ground for postulating new neighborhoods based on changing social values. Several avenues for study have been suggested

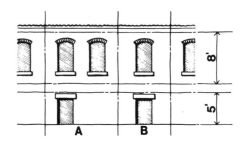

7. *Elevation of Houses in Santa Pau, Spain.* A different interpretation of land use is expressed in this town built in A.D. 1020. Animals live on the first floor (with lowered ceiling), and families above in a row house. Land-use density is much higher than in the detached single-family house. House sizes are discernable by door placements (A, B).

6. One unusual example of the trade-off game is student housing on the "Lawn" at the University of Virginia. These rooms facing the green central Lawn (or mall) of the University have no baths—students must walk outdoors a considerable distance to reach toilet and shower facilities. In addition, no parking is allowed near the rooms. There is, however, a waiting list for the rooms, honor students being given preference. This is an obvious example of sociocultural aesthetics. This older part of the University, designed by Thomas Jefferson, is considered the most desirable place to reside.

7. *House Form and Culture*, p. 133.

8. Elevation of Houses in Ainsa, Spain. Again, the private house is raised *above ground,* but in this example, the first floor becomes an arcade shop. Each house (A, B) has a different width, the only public requirement being that the owner continue the arcade. This accounts for the variety of arches.

9. Elevation of Lagny House Proposal, by Corbusier. In this example (nine centuries after Ainsa), Corbusier adds to an older concept; now the first floor is used for the automobile. Corbusier made many proposals for separating the public domain—city, cars, etc., from the private domain of the family above.

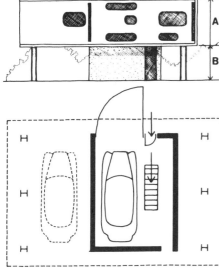

10. First Floor Plan of Lagny House Proposal. This floor plan is devoted entirely to the car.

in the illustrations. All new neighborhoods postulated could be tested using an anthrophysical approach.

Until we understand our social and political problems, and come to grips with them, we cannot expect any significant improvement of our housing environment. This is the basic weakness of the present administration's "Operation Breakthrough." Our major problem is not technological but social. There is nothing wrong with studying module units and how to mass-produce component units (although any refrigerator builder is a master in this craft), for we now need quantities of housing, and technology can help in this area. The deeper issue, however, involves value decisions. What will the union plumbers say when no plumbing is needed? If the house units are physically joined, will the public buy them? What social mix will work so that we do not simply build a slum? In my opinion the government would be better advised to spend its research monies exploring "open choice" possibilities, so that every citizen could move to suburbia, *if he desired.*

If we accept the fact that it is time to reexamine our social-political structure, we discover that among the first specific priorities for change will be property taxation. Present methods of taxation on land and home ownership must be changed. A ghetto owner can make more money by moving out, renting his

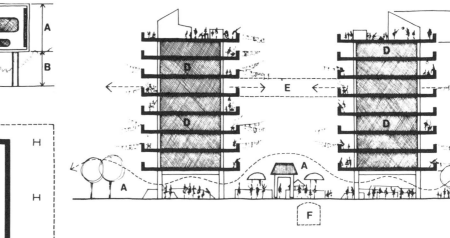

11. Section of New Apartments in Hong Kong. The design of these apartments is a logical extension of Corbusier's ideas on city planning: each family unit (D) has an outdoor balcony; the ground floor is left open (A), either for shops or as public space; and the rooftop (C) is a primary school. Many of these ideas might be developed through deed restrictions (see chapter 2, illus. 44). The next step is to connect units in the air (E), as has been done at Park Hill, Sheffield, England. Burying the rapid transit (F) completes the separation of public and private spaces.

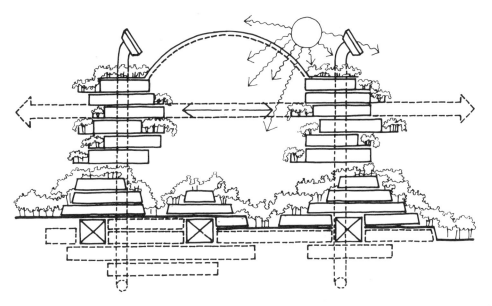

12. *A 1980 Terrace Apartment.* The final step in the evolution of the raised apartment home. Terraces provide each family with outdoor space; crosslinks tie units together; air scoops service underground rapid transit and parking; the first floors are continuous public shops, parks, etc. Skydomes filter sunlight. Proposals similiar to this have been made by the Archigram group in England and the Metabolists in Japan. The real question is not whether such a scheme is structurally and ecologically feasible (it is), but rather whether modern man will accept such a living pattern—and whether it is truly desirable.

property, and putting *no* maintenance into it, because he thus reduces its assessed value while making an inordinate profit. Two suggestions for property tax reform include:

1) No tax on first family home ownership if under a certain evaluation (say $25,000); graduated taxation above this.
2) Extra tax penalties for failure to maintain housing quality, with tax *reductions* on a graduated scale if improvements are made—this could be spread over several years (this would encourage housing rehabilitation rather than abandonment).

An equally serious area of tax reform is the control of land speculation. The purchase of a farm near a future cloverleaf intersection should not result in sudden profit either by accident, or by graft.[8] Two possibilities for study include:

1) No speculative profit unless land is held for two years.

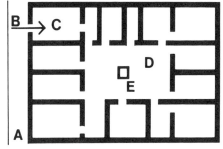

13. *The Courtyard House, Egypt.* Another basic concept of the house. Man is oriented inward to a private courtyard. In this example from 2000 B.C., no windows face the street (A), merely a single door (B), which opens to an office (C). All interior rooms face into the courtyard (D) with a central cistern well (E).

8. There are too many examples of six-month owners, doubling their investment, —this specifically includes sudden state purchases for lands needed for prisons, roadways, and postoffices.

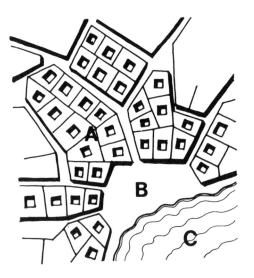

14. *Courtyard Houses, Greek Village.* In this assemblage of courtyard houses (A) all pathways lead to a small plaza (B) facing the ocean (C). The Greek islands contain hundreds of examples, such as Mikinos.

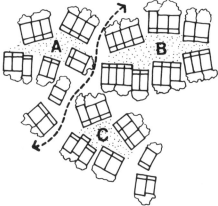

15. *Courtyard Houses, India.* This example, similar in concept to Vastrapur, is an adaptation of the courtyard house. In this case the houses are merely rooms with verandahs facing inward to a loosely organized outdoor room. These outdoor courtyard rooms are usually segregated according to caste (A, B, C).

2) After two years, a reasonable profit might be allowed tax free (say 6 percent a year), but profit above this not allowed.

These innovations along with others, (including changing the inane means of supporting schools) would begin to control land speculation and could foster the beginning of a sensible land planning policy.

In this manner social change could force taxation change, and thus allow new land-use choices to be explored. Until this point is reached, the architect-planner cannot begin to postulate real alternatives to the dreary prospects of horizonless subdivisions.

If one believes that a housing environment should be an enjoyable place, he cannot help but question what is happening to our order of national housing priorities. Below is a partial list of worthwhile goals:

To feel, again, a sense of place, with sequential movement, landmarks, nodes of interest, and ordered systems of form.

Within these ordered systems to be able to walk to school, walk to work, walk to shops—walk anywhere.

To feel safe within your neighborhood; to enjoy clean air and sunshine; to enjoy a rich mix of children, youth, and adults as friends.

To understand where your front door is; to be able to look at a tree outside your window, not a telephone pole; to have a simplified system of services (putting cars, lighting, telephones, garbage, sewage, water, and gas all in proper perspective); to have within your house both a place for television *and* a place for quiet dining with candles and wine.

And finally, to feel again that one has chosen his home on the basis of his own needs and desires.

Throughout the course of this research, and in the writing of this study, one conviction has never changed: that the architect-planner's role is to examine and understand societal forces as they apply to physical form, and then, given the opportunity, to shape our environment so that these forces are appropriately, (that is, aesthetically) reflected in the form. The anthrophysical method is intended as a tool to help in this task.

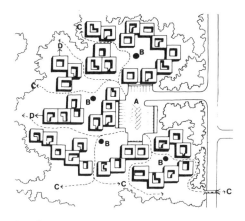

16. *Courtyard Houses, Modern Proposal.* This recent proposal suggests grouping cars in a central area (A), with courtyard houses around activity nodes (B). Many of the houses also have access to public green area (D). Pathway system (C) would include pedestrian overpass at roadway. While this proposal would obviously increase suburban densities, while also increasing open space, it is uncertain whether residents would accept the long walk from parking space to family unit.

17. *Plan for a Modern Courtyard House.* Utilizing today's technology, it is easy to construct sprayed concrete walls with irregular shapes. While the resultant architectonic form would be exciting, the question of servicing remains. This example might be tested by using "trade-off" possibilities and an anthrophysical approach.